Ordinary Places, Extraordinary Events
Citizenship, Democracy and Public Space in Latin America

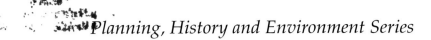

Planning, History and Environment Series

Editor:
Professor Dennis Hardy, Emeritus Professor of Urban Planning

Editorial Board:
Professor Arturo Almandoz, Universidad Simón Bolivar, Caracas, Venezuela
Professor Gregory Andrusz, London, UK
Professor Nezar AlSayyad, University of California, Berkeley, USA
Professor Robert Bruegmann, University of Illinois at Chicago, USA
Professor Meredith Clausen, University of Washington, Seattle, USA
Professor Jeffrey W. Cody, Getty Conservation Institute, Los Angeles, USA
Professor Robert Freestone, University of New South Wales, Sydney, Australia
Professor Sir Peter Hall, University College London, UK
Professor Peter Larkham, University of Central England, Birmingham, UK
Professor Anthony Sutcliffe, Nottingham, UK

Technical Editor
Ann Rudkin, Alexandrine Press, Marcham, Oxon, UK

Published titles

Of Planting and Planning: The making of British colonial cities by Robert Home

Planning Europe's Capital Cities: Aspects of nineteenth-century urban development by Thomas Hall

Politics and Preservation: A policy history of the built heritage, 1882–1996 by John Delafons

Selling Places: The marketing and promotion of towns and cities, 1850–2000 by Stephen V. Ward

Changing Suburbs: Foundation, form and function edited by Richard Harris and Peter Larkham

The Australian Metropolis: A planning history edited by Stephen Hamnett and Robert Freestone

Utopian England: Community experiments 1900–1945 by Dennis Hardy

Urban Planning in a Changing World: The twentieth century experience edited by Robert Freestone

Twentieth-Century Suburbs: A morphological approach by J.W.R. Whitehand and C.M.H. Carr

Council Housing and Culture: The history of a social experiment by Alison Ravetz

Planning Latin America's Capital Cities, 1850–1950 edited by Arturo Almandoz

Exporting American Architecture, 1870–2000 by Jeffrey W. Cody

Planning by Consent: The origins and nature of British development control by Philip Booth

The Making and Selling of Post-Mao Beijing by Anne-Marie Broudehoux

Planning Middle Eastern Cities: An urban kaleidoscope in a globalizing world edited by Yasser Elsheshtawy

Globalizing Taipei: The political economy of spatial development edited by Reginald Yin-Wang Kwok

New Urbanism and American Planning: The conflict of cultures by Emily Talen

Titles published 2006

Remaking Chinese Urban Form: Modernity, scarcity and space, 1949–2005 by Duanfang Lu

Planning Twentieth Century Capital Cities edited by David L.A. Gordon

Titles published 2007

Olympic Cities: City agendas, planning, and the world's games, 1896–2012 edited by John R. Gold and Margaret M. Gold

Planning the Megacity: Jakarta in the twentieth century by Christopher Silver

Ordinary Places, Extraordinary Events: Citizenship, Democracy and Public Space in Latin America edited by Clara Irazábal

Designing Australia's Cities: Culture, commerce and the city beautiful, 1900–1930 edited by Robert Freestone

Ordinary Places, Extraordinary Events

Citizenship, Democracy and Public Space in Latin America

edited by

Clara Irazábal

Routledge
Taylor & Francis Group

LONDON AND NEW YORK

First published in 2008
by Routledge
2 Park Square, Milton Park, Abingdon, Oxfordshire OX14 4RN

Simultaneously published in the US
by Routledge
270 Madison Avenue, New York, NY 10016

Routledge is an imprint of the Taylor & Francis Group

Typeset in Palatino and Humanist by PNR Design, Didcot
Printed and bound in Great Britain by The Cromwell Press, Trowbridge, Wiltshire

This book was commissioned and edited by Alexandrine Press, Marcham, Oxfordshire

British Library Cataloguing in Publication Data

A catalogue record of this book is available from the British Library

Library of Congress Cataloging in Publication Data
Ordinary places, extraordinary events : citizenship, democracy, and public space in Latin America / [edited by] Clara Irazábal.
   p. cm. — (Planning, history, and environment series)
Includes bibliographical references and index.
ISBN 978–0–415–35452–3 (hb : alk. paper) — ISBN 978–0–203–00121–9 (ebk)
   1. Public spaces—Political aspects—Latin America—Case studies. 2. Public spaces—Social aspects—Latin America—Case studies. 3. Cities and towns—Latin America—History. I. Irazábal, Clara.

HT127.5.O73 2008
307.76098—dc22

2007034960

ISBN10: 0–415–35452–8 (hbk)
ISBN10: 0–203–00121–4 (ebk)

ISBN13: 978–0–415–35452–3 (hbk)
ISBN13: 978–0–203–00121–9 (ebk)

Contents

Part II: Place, Citizenship and Nationhood

Preface and Acknowledgements

Revolutionary events generally take place in the street. Doesn't this show that the disorder of the street engenders another kind of order?

Henri Lefebvre

Public space and the public sphere represent conjoined arenas of social and political contest and struggle.

Setha Low and Neil Smith

From 11 to 13 April 2002, thousands of men and women took to the streets of Caracas, Venezuela, my hometown. Ordinary citizens transformed history in Venezuela through a series of extraordinary events in the public spaces of the city. The *coup d'état* which ousted Venezuelan President Hugo Chávez and the reaction to it which reinstated him were possible because of the massive street demonstrations staged by both those who opposed and supported his regime. The location and accessibility of those spaces were determinants of how these events unfolded.

History was in the making in Venezuela, but I could only experience it vicariously from my current place of residence in the United States. However, my occupation as an academic afforded me the means of coping with both my emotions and my curiosity about these events and their wider significance and to offer others something that could help them make sense of the processes at play – I decided to write about them. As I reflected on the role of public space in the extraordinary events in Venezuela, the long history of events in which public space had been essential for the defence, establishment, maintenance, or reinforcement of democracy in Latin America became evident.

Latin America is replete with examples of such events in the otherwise ordinary public spaces of their most important cities. Some of these political demonstrations have played a key role in the demise of totalitarian regimes, and/or in the subsequent re-establishment of democracy. In the light of this, I decided that an analysis of extraordinary events, public space, democracy and citizenship is best served through the study of several Latin American cities with which Caracas shared important aspects of history and culture. The idea for an edited volume thus took hold.

The collaborators to this book are a group representing various nationalities and disciplines, genders and places of residence. Many document places where they have either resided or still do so today, thus enriching the case studies with the sort of ethnographic insight that, for the most part, only a local can acquire. Not much has been written about Latin American urban studies in English, and less so by Latin Americans themselves. My hope is that this book will fill this void by bringing those voices to the foreground, allowing us the opportunity to narrate and interpret our own stories. Some of the chapters in the book were translated from Spanish and Portuguese.

I owe enormous gratitude to the contributors who enthusiastically embraced the task of providing the picture of Latin America's conditions of citizenship, democracy, and public space, which this book reveals. All were responsive to my reviews of their drafts, editorial guidance, and coordination throughout the process and for this too I am extremely grateful.

I also want to acknowledge the scholars whom I asked to provide reviews of some chapters, in particular Barbara Lynch, Rafael Pizarro, Macarena Gómez-Barris, Teresa Vázquez, and Ramzi Farhat. I am very grateful to the students and friends that helped me with translations and different editing tasks, including Joel Ulloa, Reina Rangel, Angelica Loa, Virginia Gómez-Tapia, Licínia McMorrow, Roberto Leni, Jose Prado, Cristian Pliscoff, Ramzi Farhat, and Gabriel Fumero. Very special thanks go to Ramzi Farhat for his thorough copy editing work. I also appreciate Vice Dean of Research Genevieve Giuliano for picking up the tab on some of the assistants' fees. I owe special recognition to my husband Gabriel Fumero who lovingly and patiently accompanied the project in many ways, from the design of the book cover to the intellectual encouragement of the contributors at the 2006 LASA Conference.

Finally, I dedicate this book to the late John Foley, a British urban planner who for 30 years made the ordinary places and extraordinary events of the Venzuelan people his own, and to all the people of Latin America who take their politics to the streets in hope of a better world.

References

Lefebvre, Henri (1970, 2003) *The Urban Revolution* (translated by Robert Bononno). Minneapolis, MN: University of Minnesota Press.
Low, Setha and Smith, Neil (eds.) (2006) *The Politics of Public Space*. New York, NY: Routledge.

Clara Irazábal
Los Angeles
November 2007

The Contributors

Miriam Chion, an architect and planner, is an Assistant Professor in the Department of International Development, Community, and Environment, Clark University, Worcester, MA. Her research includes the spatial capital of central cities, transformation of Metropolitan Lima in the global context, and emerging and essential industries in San Francisco.

Xóchitl Cruz-Guzmán, a sociologist, is Professor at the Universidad Autónoma Metropolitana, Mexico. She has specialized in public space and citizen participation, global cities, situational analysis, and qualitative research methods applied to Latin American urban studies.

John Foley, who sadly died in December 2006, was a British urban planner, who lived for 30 years in Venezuela, and for the last 20 was Professor in the Urban Institute at the Universidad Central de Venezuela. His research addressed the links between planning theory and communicative practice, and local social movements and their role in the conformation of urban space in the Venezuelan context.

Hans Fox Timmling is Professor in the School of Architecture, Universidad de Santiago de Chile. His research has centred on urban and regional development in Colombia and Chile. He is currently working on urban design and regional planning projects.

Robert Alexander González is an Assistant Professor in the School of Architecture, Tulane University, New Orleans, LA. He is the founding editor of the bilingual journal *Aula: Architecture & Urbanism in Las Americas*. His AIDS memorial for Key West, Florida, has received several national awards.

Clara Irazábal is an Assistant Professor of Urban Design and Planning in the School of Policy, Planning, and Development at the University of Southern California, Los Angeles. She has worked as consultant, researcher, and academic in Venezuela, Brazil, Colombia, Mexico, and the USA. Her research encompasses the politics of urban planning and design, ethnicity and space, and criticism of transnational urban design.

Susana Kaiser teaches Media Studies and Latin American Studies at the University of San Francisco, California. Her research focuses on communication, human rights, and cultural/political memory. Her book, *Postmemories of Terror: A New*

Generation Deals with the Legacy of the Dirty War explores young Argentineans' memories of the military dictatorship, including their interface with media representations of this period.

Zeuler R. Lima is an architect and Assistant Professor at the Graduate School of Architecture and Urban Design, Washington University at St. Louis, MO. His most recent project is a book about the work of Italian-Brazilian architect Lina Bo Bardi and her relationship to Brazilian modernism. As a designer, he received several awards and prizes in national design competitions and exhibitions in Brazil.

Wiley Ludeña Urquizo, an architect, teaches at Universidad Ricardo Palma, Lima and in the Postgraduate Section of the Architecture, Urbanism and Arts School, Universidad Nacional de Ingenieria, Lima, where he is the founder and Director of the Urban Renewal Masters Programme. His research is related to the themes of history and theory of urbanism, especially in Peru.

Vera M. Pallamin is Associate Professor in the School of Architecture and Urbanism, Universidade de São Paulo, Brazil. She has organized meetings and conferences dealing with urban space, culture and the public sphere and has created several public art works in São Paulo.

Alberto Saldarriaga Roa, an award-winning architect, is the Director of the Graduate Programme in History and Theory of Art and Architecture in the School of Arts, Universidad Nacional de Colombia, Bogotá. He has published several books on historical and theoretical urban and architectural issues, with emphasis on Colombian subjects.

Roberto Segre, for many years an academic in Cuba, is Emeritus Professor of the Graduate Programme in Urbanism at the School of Architecture and Urbanism, Universidade Federal do Rio de Janeiro. He has written more than 30 books and some 400 articles on the architecture and urbanism of Latin America and the Caribbean.

Sergio Tamayo is a Professor and a member of the Political Analysis Research Group in the Department of Sociology, Universidad Autónoma Metropolitana, Azcapotzalco, Mexico. He specializes in collective identities, social movements, urban identities, practices of citizenship, appropriations of public space, and qualitative research methods.

Rodrigo Vidal Rojas, Director and Professor at the School of Architecture and Director of the Master Plan of the Universidad de Santiago de Chile. His research focuses on culture and urban politics, sprawl, urban architecture, and bio-climatization of public spaces.

Prologue

Ordinary Places, Extraordinary Events in Latin America

Clara Irazábal

Scholars have argued that public space is a prerequisite for the expression, representation, preservation, and/or enhancement of democracy (Sassen, 1996; Holston, 1989, 1999; Caldeira, 2000; Low, 2000; Low and Smith, 2006). However, this optimistic outlook is betrayed in reality by the many examples in recent history when public spaces have been used for the deployment and reproduction of totalitarian regimes. In the Americas, we can recount the experiences of Pinochet's Santiago, Videla's Buenos Aires, Strossner's Asuncion, and Pérez Jiménez's Caracas, among others. Yet, even in those cases, political demonstrations in public spaces conversely played a critical role in the eventual revocation of those regimes, and/or in the subsequent re-establishment of democracy. In both the past and the present, public spaces have been privileged sites for the enactment and contestation of various stances on democracy and citizenship in the public sphere. Indeed, the *public sphere*, as the intangible realm for the expression, reproduction, and/or recreation of a society's culture and polity, usually encompasses divergent political visions and nurtures acute social confrontations which are played out in the more tangible *public space*.

This book provides a multidisciplinary approach to the study of citizenship and democracy encompassing both the notion of the public sphere and its spatiality – i.e., its re-presentation in space. Our aim is to redress an imbalance in the literature which has tended to focus on a socio-political approach at the expense of a literally more 'grounded' perspective. Thus, the case studies here help to expand and literally to ground the notion of the public sphere in the realm of physical public spaces. The uses and meanings given to and derived from public spaces in the processes of making life and making history (Flacks, 1988) have traditionally been centrepieces of both conceptual and empirical analyses of democracy and citizenship. Public space dynamics also provide ways in which to gauge, analyse, and document the value structures that lie at the core of particular societies and cultures. The selection of case studies from Latin America provides an unprecedented opportunity to look at cities with comparable cultural and political trajectories to investigate the use and meaning of particular urban spaces by ordinary people in extraordinary, history-making events. The collection encompasses multidisciplinary studies of nine Latin American major cities: Mexico City, Havana, Santo Domingo, Caracas, Bogotá, São Paulo, Lima, Santiago, and Buenos Aires.

There has been a long tradition of study of contemporary uses and meanings of urban spaces, particularly since the 1960s.[1] Yet most have focused on North American and European cases, emphasizing the analysis of everyday experiences in those places. This book makes novel contributions in two significant ways: first, by focusing on Latin American cases, which are understudied in the literature in English; and second, by emphasizing the extraordinary uses and meanings of those spaces, rather than the everyday experiences of space users. By so doing, these studies contribute to the exploration of Jürgen Habermas's two-tiered concept of society as 'lifeworld' and as 'system' (1991), shedding light on the difference, yet complementarity of the processes of 'making life' and 'making history' (Flacks, 1988), considering the latter as those enacted by ordinary people in ordinary spaces at extraordinary events.

Similarly, there are few books in English that deal explicitly with urban design and planning in Latin America.[2] This book proposes to weave multidisciplinary perspectives in exploring the dynamics of urban space in several Latin American cities in recent history. A few more studies explore the transitional dynamics of democracy and citizenship in Latin America from sociological and political sciences perspectives.[3] None, however, explicitly scrutinizes the development of democracy and citizenship in *physical* urban space, which empirically grounds these critical debates. By adopting the latter route, this book reawakens awareness of the role of space in the politics of culture and the culture of politics of cities and countries, specifically in Latin America. Lastly, this book builds upon previous theoretical works that delve into issues of the public sphere, democracy, and citizenship,[4] and contributes to these theoretical endeavours by probing the largely unexplored Latin American cities to test these ideas and further advance theory.

Conditions in the world evidence that any naïve hope in a benign globalization must be discarded. Some scholars are already theorizing what they conceive as a post-globalization era:

post global is not an end to globalization but the emergence of a different kind of engagement that is sharply at odds with the visions of liberal, multicultural globalization. Here, both religious fundamentalism and imperial hegemony begin to emerge as the new forms of global engagement. (IASTE, 2004)

We argue in this book that there is a third form of global engagement – that which (often desperately) holds on to the visions of a liberal, multicultural globalization in the politics of identity formation and/or nation building, particularly enacted in urban spaces. Invested in this third-way vision, we interrogate the role of both traditional and post-traditional Latin American urban places in this (post)glocal era, whereby the post-traditional is understood as a spatio-political repositioning that unsettles the historically developed relationships between places and meanings. The chapters help interrogate the fate of the link between public spaces and the construction of citizenship and democracy in this era, scrutinizing the various reworkings of identity, ethnicity, and other traditions of belonging. A key component addressed is the reworking of the construction of the histories of peoples and places, and their connection to and relevance in the post-traditional moment.

This interrogation makes a distinction between 'making life' and 'making history' events and processes (Flacks, 1988). Life-making processes are equivalent to everyday-life practices, the reproduction of quotidian life. When there are ruptures in ongoing life-making processes, events may become transcendental and attain a level of history-making. Habermas's notions of 'lifeworld' and 'system' are also valuable in explaining how and when disruptions of the lifeworld transform the system. He explains the colonization of the lifeworld by the system. Although acknowledging this condition and making it part of our case study analyses, we emphasize the exploration of the inverse dynamics, i.e., the colonizations of the system by the lifeworld. The lifeworld exists within the system, but when lifeworld's most fundamental rhythms are subverted, the system may experience transformation of a reformist or revolutionary nature. In a reformist transformation, the system survives, but it is (re)adjusted. In a revolutionary transformation, however, the system is replaced by a different order. A revolutionary event or process may produce a transitional systemic phase in which it may be inaccurate to talk about another system being in place. But revolutions head towards replacements of one system by another of a different nature. Historically, transitional phases from one system to another have shown varying degrees of success and have had diverse degrees of order/chaos and time lengths. System and lifeworld interact in complex and fluid ways, and paying attention to the under-explored inverse dynamic – the colonizations of the system by the lifeworld – sheds novel light on the understanding of processes of democracy and citizenship, reforms and revolutions.

Therefore, we focus particularly on the moments in which 'cracks' in the lifeworld have given way to transitions from making life processes to making history episodes – on extraordinary events in otherwise ordinary public places. Often the significance of these extraordinary events in cities which are political, economic, and/or cultural capitals have transcended their urban territories and impacted the entire country, reforming or revolutionizing its history, and in some cases, its system.

The intent in this book is thus to flip around the emphasis of most studies of urban places, which focus on the everyday life of great places. This is not to say that the latter aspect of the use of place is not discussed. Rather, it is included in an expanded scope that recognizes the wider range of opportunities of exploring conditions of citizenship and democracy in public space through the array of events that can take place there, ranging from life-making to history-making, and all the shades in-between. Thus, as subtly suggested in the cover of the book, the leading line 'Ordinary places, extraordinary events' is also meant to be read in its different permutations of words, as follows: 'Ordinary events, extraordinary places', and 'Ordinary-extraordinary, places-events'. The former alternative refers to the practices of everyday, life-making occurrences that take place in the most prominent public places. The latter suggests two conditions. It suggests the ambiguous quality of the events and places that we are qualifying, which could range from the ordinary to the extraordinary and any place in-between, both because of some of their objective characteristics, but also most importantly, because of the different interpretations that people belonging to different groups and positionalities can invest in them. The duplet 'places-events' also suggests that there are no neat boundaries between places and events (between space and time), particularly under extraordinary circumstances; that places are fluidly yet definitely marked by the (extraordinary) events that occur in them; and that events are largely shaped and sustained – i.e., embodied and made memorable – by the physical qualities of the sites in which they take *place*.

Similarly, we do not mean 'ordinary' places in the sense of pedestrian spatial quality. Rather, we mean ordinary to convey the notion that these spaces' quotidian uses support everyday, life-making practices. They are usually spaces of considerable physical dimensions and of high symbolic value and they are public – major plazas, boulevards, and streets. Conversely, by 'extraordinary' events we mean unique episodes in the life of a society in which reformists or revolutionary changes to the system occur. The extraordinary events we refer to are mainly political events, enabled by masses of people getting together in public spaces for unique purposes. In addition, some contributors offer other approaches to the understanding of extraordinary events, including historical, artistic, or economic. Even though the cases may emphasize one of such aspects, they are all multidimensional because they operate in, and help define and transform, their socio-cultural local and national contexts. Extraordinary events of all these sorts can unsettle the historically developed relationship between place and meaning, prompting collective reimagining of communities and nations and thus transforming the notions of citizenship and democracy.

Structure of the Book

Chapter 1, *Citizenship, Democracy, and Public Space in Latin America*, serves as the introduction to the topics of the book. The book is then divided into two interrelated parts. Part I, *Cities, Democracies and Powers,* discusses how central public spaces in a group of large South American cities have played significant roles in the struggles of power and redefinitions of democracy in those nations, allowing the formation, transformation, and representation of collective identities through the differential social appropriations of space.

In *Political Appropriation of Public Space,* Sergio Tamayo and Xóchitl Cruz-Guzmán analyse the political culture of participants in public demonstrations, their collective identities, and the cultural, political and social impacts of certain forms of protests and actions on Mexico City and the nation at large, through a comparative analysis of political demonstrations in Mexico City's main square, the Zócalo. The study is based on an ethnographic approach in connection to two socio-historical contexts of demonstrations held at this space. The first was carried out by the Zapatista Army of National Liberation, flanked by thousands of sympathizers in March 2001. This case is contrasted with a final electoral meeting of the winning political party during the national electoral campaign of 2000. Tamayo and Cruz-Guzmán's study underlines the processes by which urban spaces are transformed into contesting fields for, and by, different practices of citizenship, questioning the liberal and dialogical Habermasian conception of the public sphere.

Vera Pallamin and Zeuler Lima discuss transformations of São Paulo's iconic Museum of Art (MASP) and Avenida Paulista in *Reinventing the Void*. The covered space under the MASP is a hiatus in the dense Avenida Paulista – one of the most prized streets in São Paulo. Due to its location, shape, and cultural significance, the space has long been the site for events shaping both everyday life and extraordinary events. In a city of high density and few open spaces, it has provided a privileged venue for staging cultural, political, and social events. The authors explore how this void has been continuously reconstituted by both organized and spontaneous forms of public appropriation, and has thus responded to different conceptions of collective urban life. Paradoxically, the effervescence of all these events has not resulted in greater openness in public space. Today, the space in front of the MASP is still used for protests, while the space under it is restricted to private events. Pallamin and Lima point out that this disjuncture opens new meanings and new forms of contestation and appropriation of urban open spaces in the city, a process with no foreseeable closure, since different forms of strategic exclusion continuously face tactical events of resistance.

Some urban spaces do not lose strong collective memories associated with them even when they undergo significant transformation. Rodrigo Vidal and Hans Fox, in *A Memorable Public Space*, reflect on the historical meanings and uses of the square of Grand Central Station and the campus of the University of Santiago, in Santiago de Chile. By the early 1980s, the site had become one of the essential places of protest against Chile's military regime, and a bastion of opposition against the police. With the establishment of democracy, different struggles surfaced. Today, new commercial activities are bringing vitality to the area, even though most of the buildings manifest symptoms of decay. Vidal and Fox unveil the elements of the urban memory related to this place and the conditions that have been crucial in constructing its social meaning.

In *Lima's Historic Centre*, Miriam Chion and Wiley Ludeña analyse the Historic Centre of Lima, with its central role in both local and national histories throughout the city's life. This space creates a sense of place and social belonging for Lima's population, which contrasts with the increasing homogenization of the city by shopping malls and standardized office buildings. In the re-emerging role of Lima's Historic Centre as a place for the reformulations of social identity, multiple urban actors compete for the use of the space. The Centre as a place for the reworking of social identity has a particular connotation in an era of globalization, when many financial and informational transactions are placeless, i.e., occur in the space of flows. Also, many places of consumption are increasingly similar across regions and have little local identity. Places of identity, therefore, can define the makeup of a city and the imagination of a nation. Chion and Ludeña explore these processes through the novel concept of spatial capital.

Alberto Saldarriaga focuses on a place that has held many history-making, extraordinary events – *The Plaza de Bolívar of Bogotá*. The plaza started out as a typical space at the centre of Spanish urban settlements in the Americas. As an open void in a quadrangular grid of streets, it was a place for many events – religious celebrations, political demonstrations, markets, bullfights, and public feasts. This plaza has played a prominent role in Colombia's political life. Saldarriaga analyses the relationship between the physical design of the plaza and some major political events of the twentieth and twenty-first centuries. For him, the new century seems to intensify the use of the plaza for extraordinary events. These events highlight the tensions that Colombia experiences, and consequently, the socio-political and spatial renegotiations in relation to different notions of nation, democracy and citizenship that are constructed and debated among different groups in the plaza.

Part II, *Place, Citizenship and Nationhood* discusses the formation and representation of competing ideas of citizenships and nationhood through the appropriations and representations of public space by distinct groups engaged in social, cultural, and political struggles and projects. Through the lens of Caracas in *Space, Revolution and Resistance*, Clara Irazábal and John Foley examine the role of architecture and urban space since the beginning of the political transformations before and during the April 2002 *coup d'état* that forced President Chávez out of office, the subsequent *contra-coup d'état* that re-established him in power, and the aftereffects of those events. During those events, the public buildings which are the seat of political power and the military, and the private buildings which are the headquarters of the media and other protagonists, coupled with other key urban spaces, became privileged sites for the public enactment of discontent as well as staging grounds for claims of various reformulations of citizenship. According to Irazábal and Foley, the analysis is evidence that, in many ways, urban spaces and architecture were and continue to be key ingredients in these unprecedented national political disputes in Venezuela, enabling social inequalities and polarization in the country to find their expression and contestation in the capital's urban space.

In *The Struggle for Urban Territories*, Susana Kaiser discusses the key role that the creative, forceful, and disruptive public presence of activists played in shaping policies regarding memory, accountability, social justice and democratization in Buenos Aires. She starts with an overview of the Mothers of the Plaza de Mayo's communication strategies to denounce state terrorism and demand accountability. By transforming motherhood into a public activity, the Mothers pioneered the redefinition of what is public in Argentina, which is at the core of the country's human rights' struggles. By conquering physical and metaphorical territories, they shaped the style and the scope for human rights activism. Kaiser then follows with an analysis of the *escraches* organized by HIJOS – the children of disappeared people – and then focuses on recent street demonstrations and the new *escraches*, demonstrating how these strategies have been co-opted and adapted for a variety of causes. Kaiser conceives the streets of Buenos Aires as arenas of power struggles for the rewriting of memory and history.

Emphasizing the linkages between ideology and urban space is Roberto Segre's message in *Iconic Voids and Social Identity in a Polycentric City*. In Havana, extraordinary events have taken place in different spaces at different times, because the city has been continually transmuting throughout its almost five centuries of history. Segre charts the evolution of Havana's public spaces through a historical account of the eras of neoclassicism, modernity, and revolution in the city and country. He discusses some milestone events that occurred in Havana's formal and informal spaces, unveiling its urban history as impregnated with ideological conditioning that has defined people's symbolic and physical appropriations of space. Finally in 1959, the monumental Civic Plaza transformed into the pulsating Plaza of the Revolution, becoming a stage to masses of Cubans over almost half a century of subsequent communist regime. Thus, through its urban spaces, Segre evidences that Havana has vividly embodied the historic ups and downs of the conflicting ideologies that have defined the destiny of the nation.

In *Unresolved Public Expressions of Anti-Trujilloism in Santo Domingo*, Robert González interprets monuments as extraordinary events and thus expands the exploration of socio-political constructions of conflicting notions of memory and nationhood. González focuses on several anti- and pro-Trujillo monuments typically left out of tourist maps, and the national debates surrounding them. The city offers a prominent example of a public space that was used to celebrate, at one time, the aggrandizement of a tyrannical figure, and later, the reclamation of human rights: the main plaza of the 1955 *Feria de la Paz y Confraternidad del Mundo Libre*. Built by Trujillo as a world's fair, it was symbolically transformed after his assassination into *El Centro de Los Héroes*, honouring executed Dominicans who tried to overthrow his regime. In addition to this contested space, following Trujillo's assassination, innumerable spatial references to his supposed heroism and generosity were removed from the city. While all the monuments both represent and constitute extraordinary events, González claims that the monuments' invisibility from the tourist imaginary reflects a country that is still grappling to come to consensual terms with its national history.

With its array of case studies and theoretical perspectives, we believe this book will appeal not only to those interested in Latin American studies, but to anyone with an interest in cities, their peoples, politics, architecture, design and planning. Finally, we hope that the book's stories will provide inspiration to community leaders and city residents to think of their cities as laboratories for emergent citizenships.

Notes

1.   See among many other significant authors, Jacobs (1961) and Galhaguer (1993) in journalism; Whyte (1980, 1988) and Jacobs (1985) in urban planning; Hall (1959, 1966), Spradley and McCurdy (1972), Low (1999, 2000), and Low and Smith (2006) in environmental psychology, cultural anthropology and geography; Harvey (1989) in geography; Davis (1992) and Soja (2000) in urban studies; Lynch (1961) and Loukaitou-Sideris and Banerjee (1998) in urban design.
2.   Fortunately, the interest in this area is rapidly expanding. Recently, from the perspective of environmental psychology and anthropology, Low (2000) approximates the intent of this book's chapters by accounting for the history, use, and meaning of two plazas in San José, Costa Rica. Almandoz (2002) and del Rio and Siembieda's (forthcoming), respectively deal with a period of the past, and a focused study of a country – Brazil. Irazábal (2005) compares the praised city of Curbita to the similarly appreciated model

in the United States city of Portland, OR. Scarpaci (2005) ambitiously evaluates the impact of heritage and globalization politics in nine historic centres in the region, revealing an ambivalent picture regarding the cities' responses to global pressures. Lastly, Herzog (2006) explores central public spaces in Spanish, Mexican, and Mexican-American border cities, discussing their design and politics as well as broader global implications.

3. Among them, Agüero and Stark (1998); Garretón and Newman (2001); Lievesley (1999); Mainwaring and Valenzuela (1998).

4. I refer to books such as Hénaff and Strong's (eds.) (2001); Habermas (1991); Holub (1991); McGuigan (1996); Mullard (2003); Trend (1996); and Vandenberg's (2000). Similarly, exploring issues of citizenship in specific urban contexts around the world, recent books include Boudreau (2000); Holston (1999); Isin (2000); and Isin and Wood (1999).

References

Agüero, Felipe and Stark, Jeffrey (eds.) (1998) *Fault Lines of Democracy in Post-transition Latin America*. Boulder, CO: Lynne Rienner Publishers.

Almandoz, Arturo (ed.) (2002) *Planning Latin America's Capital Cities, 1850–1950*. London: Routledge.

Avritzer, Leonardo (2002) *Democracy and the Public Space in Latin America*. Princeton, NJ: Princeton University Press.

Boudreau, Julie-Anne (2000) *The Megacity Saga: Democracy and Citizenship in this Global Age*. Montreal: Black Rose Books.

Caldeira, Teresa (2000) *City of Walls: Crime, Segregation, and Citizenship in São Paulo*. Berkeley, CA: University of California Press.

Calhoun, Craig (1992, 1999) *Habermas and the Public Sphere*. Cambridge, MA: MIT Press.

Davis, Mike (1992) *City of Quartz: Excavating the Future in Los Angeles*. New York. NY: Vintage Books.

del Rio, Vicente and Siembieda, William (eds.) (forthcoming) *Contemporary Urban Design in Brazil* .

Flacks, Richard (1988) *Making History: The Radical Tradition in American Life*. New York, NY: Columbia University Press.

Galhaguer, Winifred (1993) *The Power of Place: How Our Surroundings Shape Our Thoughts, Emotions, and Actions*. New York: Poseidon Press.

Garretón, Manuel Antonio and Newman, Edward (eds.) (2001) *Democracy in Latin America: (Re)constructing Political Society*. Chapel Hill, NC: University of North Carolina Press.

Habermas, Jürgen (1991) *The Structural Transformation of the Public Sphere: An Inquiry into a Category of Bourgeois Society.* Cambridge, MA: MIT Press.

Hall, Edward T. (1959) *The Silent Language*. New York, NY: Doubleday.

Hall, Edward T. (1966) *The Hidden Dimension*. New York, NY: Doubleday.

Harvey, David (1989) *The Condition of Postmodernity: An Enquiry into the Origins of Cultural Change*. Oxford: Blackwell.

Hénaff, Marcel and Strong, Tracy (eds.) (2001) *Public Space and Democracy*. Minneapolis, MN: University of Minnesota Press.

Herzog, Lawrence (2006) *Return to the Center: Culture, Public Space, and City-Building in a Global Era*. Austin, TX: University of Texas Press.

Holston, James (1989) *The Modernist City: An Anthropological Critique of Brasília*. Chicago, IL: University of Chicago Press.

Holston, James (ed.) (1999) *Cities and Citizenship*. Durham, NC: Duke University Press.

Holub, Robert (1991) *Jürgen Habermas: Critic in the Public Sphere*. New York, NY: Routledge.

IASTE (2004) Post Traditional Environment in a Post Global World. CFP 9th Conference, Sharjah/Dubai, December.

Irazábal, Clara (2005) *City Making and Urban Governance in the Americas: Curitiba and Portland*. Aldershot: Ashgate.

Isin, Engin (ed.) (2000) *Democracy, Citizenship, and the Global City.* London: Routledge.

Isin, Engin and Wood, Patricia K. (1999) *Citizenship and Identity.* Thousand Oaks, CA: Sage.

Jacobs, Allan (1985) *Looking at Cities.* Cambridge, MA: Harvard University Press.

Jacobs, Jane (1961) *The Death and Life of Great American Cities.* New York, NY: Random House.

Lievesley, Geraldine (1999) *Democracy in Latin America: Mobilization, Power, and the Search for a New Politics.* Manchester: Manchester University Press.

Loukaitou-Sideris, Anastasia and Banerjee, Tridib (1998) *Urban Design Downtown.* Berkeley, CA: University of California Press.

Low, Setha (1999) *Theorizing the City: The New Urban Anthropology Reader.* New Brunswick, NJ: Rutgers University Press.

Low, Setha (2000) *On the Plaza: The Politics of Public Space and Culture.* Austin, TX: University of Texas Press.

Low, Setha and Smith, Neil (2006) *The Politics of Public Space.* New York, NY: Routledge.

Lynch, Kevin (1961) *The Image of the City.* Cambridge, MA: MIT Press.

Mainwaring, Scott and Valenzuela, Arturo (eds.) (1998) *Politics, Society, and Democracy: Latin America.* Boulder, CO: Westview Press.

McGuigan, Jim (1996) *Culture and the Public Sphere.* London: Routledge.

Mullard, Maurice (2003) *Democracy, Citizenship and Globalization.* New York, NY: Nova Science Pub.

Sassen, Saskia (1996) *Losing Control? Sovereignty in an Age of Globalization.* New York, NY: Columbia University Press.

Scarpaci, Joseph (2005) *Plazas and Barrios: Heritage Tourism and Globalization in the Latin American Centro Histórico.* Tucson, AZ: University of Arizona Press.

Soja, Edward (2000) *Postmetropolis: Critical Studies of Cities and Regions.* Oxford: Blackwell.

Spradley, James and McCurdy, David (1972) *The Cultural Experience: Ethnography in Complex Society.* Chicago, IL: Science Research Associates.

Trend, David (ed.) (1996) *Radical Democracy: Identity, Citizenship, and the State.* New York, NY: Routledge.

Vandenberg, Andrew (ed.) (2000) *Citizenship and Democracy in a Global Era.* New York, NY: St. Martin's Press.

Whyte, William H. (1980) *The Social Life of Small Urban Spaces.* Washington, DC: Conservation Foundation.

Whyte, William H. (1988) *City: Rediscovering the Center.* New York, NY: Doubleday.

Chapter 1

Citizenship, Democracy, and Public Space in Latin America

Clara Irazábal

It may even be, after all, that there is something irreducible and nontransferable, necessary but not quite sufficient, about the city's public street and square for the realization of a meaningfully democratic citizenship. If we support the latter, we may have to do much more to defend the former.

Holston and Appadurai (1999, p. 16)

Cautionary Tales from the Urban South: Latin American 'Grand Refusal'

How is Latin America's politics changing, and what is the role of public space in these transformations? How are people in Latin American countries expressing both their discontent with unrepresentative national regimes and also with the neoliberal agenda, which often is imposed 'from above' by multinational institutions and encouraged by the United States? Within this context, how do urban street politics transform local and national politics and relations with the USA and the rest of the world? Taking to the streets in Latin America today is a response to international issues (for example, protests against World Bank or Monetary Fund policies, or the presence of US dignitaries) and domestic ones (such as the *escraches* against torturers in Buenos Aires, or demonstrations in support or opposition to Chávez in Caracas). To analyse these processes, I employ Alain Touraine's concept of *'grand refusal'*, in which he refers to the reaction of the masses in social movements to the oppressive economic conditions caused by global neoliberalism. A 'grand refusal', however, can be more than a reaction, and can catalyse a vision for alternative socio-political projects.

Politicians critical of US foreign policy and the ensuing reign of neoliberalism are once more to the fore. In Venezuela, gubernatorial and mayoral elections favoured *Chavistas* – supporters of president Hugo Chavez in 2005, and re-elected Chavez as President on 3 December 2006. In October 2004, the left won the presidency in Uruguay – with Tabaré Vázquez – for the first time. The left had further success in the mayoral elections of May 2005. For the first time ever, eight leftist mayors took office in July 2005. In Brazil, President Luiz Inácio (Lula) Da Silva was re-elected in October 2006 and the Labour Party also gained ground in regional and local elections. In Chile, the leftist Michele Bachelet won the presidency in January 2006, Daniel Ortega in Nicaragua and Rafael Correa in Ecuador in November 2006, while in Peru the indigenous leftist leader Ollanta Humala was a serious run-off election contender in June 2006. Nine out of twelve countries in South America are now ruled by leftists, with the addition of Cuba, Nicaragua, Panama, the Dominican Republic, and Costa Rica, which have left-of-centre presidents.

In most cases, taking to the street was a crucial political strategy. Recent events in Bolivia's capital, La Paz, also deserve attention. Since 2003 sustained street demonstrations have been significant in ousting two presidents from power. As Gamarra concedes, '[t]he notion of governing from the streets is very, very prevalent in Bolivia' (cited in McDonnell, 2005, p. A3). Through street protests, road closures and strikes, indigenous and poor people demanded attention to their plight and opposed the privatization of the country's natural resources. Street politics have affected ballot box politics, as demonstrated in the elections of 18 December 2005, which resulted in a landslide triumph for the indigenous leader Evo Morales. He joined the ranks of leftist leaders Hugo Chávez in Venezuela, Luiz Inácio (Lula) Da Silva in Brazil, and Néstor Kirchner in Argentina in opposing neoliberal dictates from Washington and multinational interests.

In Latin America, the International Monetary Fund and other global organizations have exerted great pressure for the adoption of a neoliberal agenda. While arguably some reforms were necessary and benefited some sectors of the population, others have had dramatically negative consequences. In Latin America, social inequalities are among the most extreme in the world. The richest tenth of the estimated 559 million people in the continent in 2005 earned 48 per cent of the total income, while the poorest tenth earned only 1.6 per cent. These inequalities are racially and ethnically biased, with indigenous and Afro-descended peoples at a considerable disadvantage. The disparities are also clearly evident in the polarization of urban space and the existence of slums (UN-Habitat, 2005, p. 111). The rate of population growth and pace of urbanization have both increased rapidly. Latin America's population tripled in the 50 years to 2000, reaching 519 million. During the same period, the urban population grew fivefold. In 2001, 32 per cent were living in slums, more in South America (35.5 per cent), and several major cities with a much higher percentage. The structural transformation in the region's economies was instrumental in accentuating the social and spatial polarization in Latin American cities.

But those conditions are being challenged in unprecedented ways, with social groups reconstituting citizenship by reterritorializing public space. New geographies of race, class, political consciousness, and political affiliation are transforming power, knowledge, subjectivities, and ultimately, space. Significantly, the process goes both ways – transformations of space cause transformations of power, knowledge, and subjectivities. These social mobilizations continue to be propelled to a great extent by reactions to neoliberalism as disenfranchised masses demand alternative models of development. The organization, focus, and political repertoire of social movements in Latin America have changed as the eras of military and oligarchic rule ended (Foweraker, 2005). The expanding repertoire of political action includes, but is not limited to, meetings, rallies, demonstrations, concerts and performances, strikes, barricades, sit-ins, *cacerolazos*, *escraches*, and media events of all sorts. Many actions are motivated by material demands, but are often transformed or expanded into claims of civil, political, human, and cultural rights.

Opinions are mixed regarding the impact of such movements and actions in a context where every human, social, and political right has had to be won through social and political struggle against democratic regimes of 'low-intensity citizenship' (O'Donnell, 1999; cited in Foweraker, 2005, p. 123). At the height of the neoliberal era of the 1990s in which the emphasis was on economic rather than political or social development, some Latin Americanists assessed that it was 'impossible to mobilize and press for effective rights of citizenship, or strive to hold newly democratic governments to account' (*Ibid.*, p. 130). However, 'a historical perspective shows that social mobilization, whether in Latin America or elsewhere, always occurs in waves' (*Ibid.*, p. 133). Accordingly, today several Latin American countries are arguably entering the era of leftist post-neoliberal regimes with a consequent heightened use of public space for both everyday and extraordinary events, all of which grounds my claim about a new wave of social mobilization *à la* Alain Touraine's 'grand refusal'.

This charged use of public space for political protests, however, is not restricted to Latin America. Around the world, the World Trade Organization (WTO), International Monetary Fund, World Bank, the Group of 8, and the European Union have had to deal with protest during summits. Taking to the streets during the WTO meeting in Hong Kong in December 2005 had been preceded by similar demonstrations in Seattle, USA, Cancun, Mexico, and elsewhere. With unprecedented world-wide coordination, on 15 February 2003, more than 30 million people in 600 cities and around the world marched for peace and against the war in Iraq. '[T]he world witnessed the largest coordinated protests in history... [O]rdinary people the world over took to the streets to assure that their voices were heard and their sheer numbers seen' (Mitchell and Staeheli, 2005, p. 796).

There has also been a steady shift from reactive to proactive demonstrations, mainly represented by the World Social Forum (WSF). The forum meetings have become an important venue for trade unions, women's groups, and peasants' and environmental movements from around the world to learn and share organizing strategies, canvas support, coordinate world campaigns, and build alliances around a platform of justice. In this 'movement of movements' the different organizations attempt to work through the conflicts between reality and utopia, 'between real achievement and contestation of the official notion of the real' (Ruggiero, 2005, p. 297).

Ordinary Places, Extraordinary Events

In Latin America, cities are crucial to the negotiation of citizenship and governance. From celebrations and affirmations, to protests and violent acts, the case studies in this book illustrate the expanded terrain of citizenship practices challenging the 'post-justice city' (Mitchell, 2001) and exploring alternative models of development and urban solidarity. In times of crisis, and also during extraordinary collective celebrations, it is common for the population to mobilize in public spaces. Social groups and *ad hoc* collectives have taken to the streets in response to the privatization of energy resources and primary sources of employment, the globalization of commerce and communication, the politics of austerity and inflation, the degradation of urban and regional infrastructure, unsatisfactory urban services and investment in education, and paucity of jobs. Identity politics – issues of legal status, gender, sexuality, race and ethnicity – are also increasingly played out in public space. In this sense, many authors concur that '[t]he most sensible and dramatic scenario for the struggle between the neoliberal globalization and the defense of the local is the city... The city is the spatial articulation of this dispute in a world of generalized urbanization' (Cantú Chapa, 2005, p. 28).

The need to respond collectively to contemporary urban problems and to defend the right to express identity have transformed Latin American capitals from 'revanchist cities' (Smith, 1996) to 'contesting cities' (*ciudades contestatarias*) (*Ibid.*, p. 100) or to paraphrase Holston and Appadurai, 'insurgent cities'. The Zócalo and the streets of the historical centre of Mexico City are a good example. Since the neoliberal transformation of the economy in Mexico in the early 1980s demonstrations have been an almost daily occurrence. In 1995, there were on average seven demonstrations daily, and 10.4 in 1996, 70 per cent of which were organized by groups arriving in the capital from the interior (Cantú Chapa, 2005, p. 101).

Across Latin America, the transformation of the physical landscape is evident in new gated communities, new global architecture, the privatization and gentrification of older districts, and the creation of new ghettoes and edge cities (Borja, 2003a; Borja and Muxí, 2002; Irazábal, 2005). In her treatise on contemporary Buenos Aires, Zaida Muxí describes it as a 'gapped city' (Muxí, 2004, p. 163), designed with 'the strategy of fragmentation', which 'observes reality with a zoom – cutting, isolating, and resolving in a piece-meal fashion – not looking for connections' (*Ibid.*, p. 165). This concept is akin to that of 'splintered urbanism', discussed by Graham and Marvin (2001). Take the example of historic preservation in Mexico City, where Capron and Monnet (2003) expose how seemingly progressive gestures paradoxically exacerbate political, social, and economic inequalities. These findings are further elaborated on by Scarpaci (2005), who found that public-private partnerships, centralized planning, and globalization conditioned historic centre revitalization, including in Havana, favouring private commercial and tourist development and gentrification over affordable housing (the sole exception in this study is Trinidad, Cuba). The cases of Lima and Havana in this book provide evidence of some of these realities.

The spatial barriers resulting from these processes – the lack of public space, or its reduction, privatization or over-regulation (through restrictions on activities and access); the lack of access due to land use regulations, the shape of the urban grid, or availability of transport – can significantly hamper the practice of citizenship and democracy. However the processes leading to these spatial conditions do not go uncontested. Many authors, who agree that public space is essential to the maintenance of democracy in making it possible to publicize dissent, also recognize that its privatization has potentially negative political ramifications (Zukin, 1991; Sorkin, 1992; Kohn, 2004). Or as McBride (2005, p. 1002) says, 'When we lose public space, we lose democracy'. Contributing to the privatization of the urban landscape in both North and South America is the reorganization of common space in the service of consumption, the creation of new layers of undemocratic governance – for example, through Development Districts and Home Owners Associations – and jurisprudence leaning in favour of private interests (Kohn, 2004).

Rosenthal (2000) claims that the process of decline of public space in Latin America has not been as pronounced as in the United States since World War II as cars, skyscrapers, suburbanization, television, and consumerism are less prevalent, while widely used public transport systems, the interest of elites in preserving downtowns, café-oriented societies, and nationalist memory processes that valorize public places are mitigating factors. Notwithstanding these assurances, the stakes are high, and having access to public places in which people can exercise freedom of speech and relate to other social groups with shared interests is considered a prerequisite condition for democracy (Low, 2000; Low and Smith, 2006). However, taking to the streets cannot be romanticized as a panacea for all grievances or as resulting in the enactment of just laws and policies. On the contrary, street politics is often the last recourse after all formal claims against injustice have failed. However, we do not want to overplay the role of street politics. Although they often have measurable impact, public demonstrations are sometimes the last resort in an ongoing struggle against inequality. Their effectiveness in ameliorating injustice varies with the power of demonstrating groups *vis à vis* power holders, the commitment the latter have to issues of social justice and democracy, and the material and non-material resources available to respond to people's claims. Paradoxically, sometimes achieving a positive result, however partial, can void a social movement of its power and may result in the abandonment of the public space as a fruitful and dynamic arena of the political public sphere.

The alternative to the 'invited spaces' of citizenship is the 'invented spaces' of citizenship, informally created by the people, which can vary in character (Miraftab, 2004; Miraftab and Wills, 2005). Here, we advance the notion that extraordinary events in public spaces have the potential, under certain circumstances, to expand dramatically invented spaces of citizenship. Evidently, 'peaceful negotiations and clever, persuasive tactics are not always effective at expanding the spaces of citizenship practice' (Miraftab and Wills, 2005, p. 208). In effect, most of these demonstrations in public spaces do not cause radical transformation, and many go almost unnoticed. But there are a few that result in radical transformation, and sometimes it is the cumulative effect of several or even many that bring about significant change. This book bears testimony to all these different possible scenarios. The effectiveness of street action is also limited if 'street fatigue' ensues when sustained mobilization is without proportional gain in the political arena. Such was the case of the opposition to Chávez in Venezuela. After taking their politics to the streets of Caracas for years against Chávez's regime to no avail, maintaining the level of mass mobilizations sustained earlier became impossible.

Citizenship, Democracy and Public Space in Latin America

Citizenship and Democracy

Theories of citizenship

Thomas H. Marshal (1964) defines citizenship as 'a status bestowed on those who are full members of a community. All who possess the status are equal with respect to the rights and duties with which the status is endowed' (Marshal quoted in Friedmann, 2002, p. 168). The story of the progression from civil rights in the eighteenth century, to political rights in the nineteenth century, and finally social rights in the twentieth century typifies the omission of the domain of culture from citizenship. In the contemporary view of citizenship, cultural rights 'are important in expanding the legal framework of governance into the cultural sphere, but the main issues are less normative than symbolic and cognitive, since it is about the construction of cultural discourses' (Delanty, 2002).

Here, we employ citizenship as an analytical tool to scrutinize the relationships between different social groups, and between the state and civil society. Citizenship also allows the scrutiny of the struggles for the expansions of social, cultural and political rights, the dynamics of identity politics, and the disputes over meanings and practices of participation. It is a framework to understand the existence and distribution of resources, the interactions between the public and the private, the social and the individual, and the modern and the traditional (Tamayo, 2004). With all the demographic, social, political, economic, and cultural restructuring in the world today, it is no longer adequate to restrict membership in society within the frame of the nation-state. Traditional, nation-derived notions of citizenship have suppressed difference. But the scales, institutions, and spaces of citizenship are morphing, and new understandings invite us to think in 'thick', multi-layered concepts (Yuval-Davis, 1999). Citizenship has been expanded to encompass cultural claims, human and local rights, and significantly, the rights to the city. Theories of radical democracy and planning, cultural and sustainable citizenship, and social justice do inform rights claims as groups enact class, gender, ethnic and racial, immigrant, religious, and sexual orientation identity politics in public spaces (Boudreau, 2000).

Cultural citizenship, inclusive citizenship (Gaventa, 2002), active citizenship (Kearns, 1995), and insurgent citizenship (Holston, 1995) all redefine the practices, values, and rules of society today. Lee's 'performative paradox' (Lee, 1998) of the gulf between citizenship as defined by law and as enacted subversively in practice is also captured by Boudreau in her 'performative citizenship' whereby '[t]he right to inhabit the city, to be there, to be politically active regardless of one's legal status, and the right to claim rights, are written nowhere in constitutional definitions of citizenship' (Boudreau, 2000, p. 132). However new Latin American constitutions, such as in Brazil and Venezuela, are radically changing this reality. Winocur's four perspectives on the transformation of citizenship help us make sense of the change (Winocur, 2003). First, the notion of citizenship is redefined to include the right to be different from the dominant national community, and citizenship is understood as fluid and dynamic with rights and values constructed through practices and discourses. Cultural citizenship thus can be seen as an extension of Marshall's progression from civic and political to social citizenship. And in the context of globalization, it is evidently a form of citizenship that extends beyond nationality (Delanty, 2002). The gender-based events in São Paulo and the artistic events in Bogotá discussed in this book are examples of extraordinary events mobilized to claim the 'right to be different'.

Secondly, citizenship is transformed as people's interests shift from the political to the social. Put in another way, experience, more so than formal institutional politics, organizes people's identities. This 'recuperation of politics as an inherent capacity of citizens' (Lechner, 2000, p. 31) is evidenced in the *Madres* and *Hijos* movements in Buenos Aires. Thirdly, the notion of citizenship is directly linked to rights of minorities and marginalized groups to quality public space. In this view, people become citizens through their participation in the conception, construction, and management of the city, and particularly, through the negotiations of the use of public space. These dynamics are evidenced for instance in the reterritorializations of public spaces in Lima and Caracas. Lastly, the forth perspective on citizenship explores the impact of communication media on participation in the public sphere, the formation of public opinion and values, the sense of societal belonging, and ultimately, the appreciation and use of public spaces. Its impact has been felt throughout, most dramatically in the case of Venezuela. In Buenos Aires, a combination of media – graffiti, flyers, and Internet – have helped Hijos uncover the identity of former torturers. In Rio de Janeiro, TV serves both to unite Cariocas during certain festivities and to terrorize them when drug gangs seize the city, while in Mexico City, reports in the media can make or break political candidates and campaigns.

When the effects of the media and public space are factored in, citizenship becomes less an abstract notion of political rights and duties in a nation-state and more a flexible notion that is 'popular, polysemiotic, and instrumental' (Winocur, 2003, p. 248). The media play a very critical role in the construction of citizen identities, and can significantly legitimize or delegitimize citizen practices in public space, as claims can be presented as 'rightful' or criminalized as 'inauthentic' (Miraftab and Wills, 2005). In particular cases, the manipulated portrayal in the media of public demonstrations, marches, strikes, riots, coups and contra-coups, and exit polls has effectively constructed 'virtual geographies' (Crang *et al.*, 1999; Wark, 1994), leading to 'imagined communities' (Anderson, 1983) and 'imagined geographies' (Gregory, 1994).

The city as the site of insurgent citizenship

Cities are increasingly functioning as a privileged locus for the formation of new claims to citizenship (Sassen, 1996). The concepts of insurgent urbanism and insurgent citizenship, introduced by Holston (1995) and further articulated by Sandercock (1998*a*), Friedmann (2002), Miraftab (2004), and Miraftab and Wills (2005), provide the means with which to analyse these phenomena. Holston and Appadurai (1996, p. 50), largely referring to squatter settlements and labour or homeless camps, define spaces of insurgent citizenship as 'situations which engage, in practice, the problematic nature of belonging to society'. In furthering the notion of insurgent citizenship, this book argues that extraordinary, short-lived events in urban public places can nevertheless have a lasting impact and a transformational effect on cities and nations, and thus constitute spaces of insurgent citizenship.

Insurgent citizenship challenges and problematizes the normative basis of citizenship in capitalist societies. According to Miraftab and Wills (2005, p. 202), it 'challenges the hypocrisy of neoliberalism: an ideology that claims to equalize through the promotion of formal political and civil rights yet, through its privatization of life spaces, criminalizes citizens based on their consumption abilities'. This 'consumerist citizenship', promoted through the privatization of open space, the creation of gated communities and privatized edge cities, the criminalizing of homeless and immigrants, and the disciplining of insurgent groups (MacLeod, nd), defines what Mitchell has labelled the era of the 'post-justice city' (Mitchell, 2001), in which citizenship rights are taken away from the ones who cannot particiate in the neoliberal economy. To counteract this trend, we adopt Miraftab's (2004) call for the recognition of the improvised, invented spaces of citizenship:

'Invited' spaces are defined as the ones occupied by those grassroots and their allied non-governmental organizations that are legitimized by donors and government interventions. 'Invented' spaces are those, also occupied by the grassroots and claimed by their collective action, but directly confronting the authorities and the status quo. While the former grassroots actions are geared mostly toward providing the poor with coping mechanisms and propositions to support survival of their informal membership, the grassroots activity of the latter challenges the status quo in the hope of larger societal change and resistance to the dominant power relations. (Miraftab, 2004, p. 1).

Here we emphasize the 'invented' spaces of citizenship created, used, and appropriated by people where recourse to 'invited' spaces is ineffective. In developing countries, the negative externalities of globalization and neoliberalism have been felt more poignantly (Kabeer, 2002). Thus, it is particularly the poor who are at the centre of the drama of evolving forms of citizenship, mobilizing to attain rights to the city (Lefebvre, 1996). Frequently these contesting dramas are performed in the streets, expanding the 'invited' public sphere and creating new spaces and practices (Isin, 1999; Rose, 2000). We also second the feminist critique of liberal notions of citizenship that assume the identities, rights, and duties of citizens as fixed and universal (Fraser and Gordon, 1994; Roy, 2001, 2003; Sandercock, 1998*b*; Wekerle, 2000; Young, 1990; Yuval-Davis, 1999; Miraftab 2004; Miraftab and Wills, 2005).

Space and the dynamic spatio-temporal scales of citizenship

Historically, citizenship marked the state of belonging and commitment to a specific place (a city-state or borough), with rights and duties performed in this context (Isin, 2002*b*). This place-rootedness of citizenship was somewhat diluted by the geographic expansion to national citizenship from the late eighteenth century onwards, but recent transformations in government and the saliency of cities as economic and cultural engines have arguably strengthened the previous connection: active citizens act for and within place-based communities. In the context of governance, decentralization – the growing mode of 'governing through communities' – shifts the emphasis from 'national citizens' to the practice of responsibilities by 'active citizens' in sub-national communities (Rose, 2000).

As Bullen and Whitehead (2005, p. 500) argue, it is the inclusion of space in the analysis of citizenship – particularly through the excavation of 'heterotopias, post-modern places, and closet spaces of citizenship' – that has helped reveal 'a whole range of citizens and modes of radical/alternative citizenship forged around issues of gender, sexuality, ethnicity, age, class and religion, which had previously been excluded from analyses of citizenship'. Massey (2004), Amin (2004), and Desforges *et al.* (2005), among others, advocate 'a new "politics [and citizenship] of propinquity"' with a focus on diversity within places (Amin, 2004, p. 38). Similarly, Massey (2004) describes a form of spatially bounded citizenship that is based on continual, and sometimes conflict-ridden negotiation. With increased mobility and the rise of trans-nationalist identification, new global forms of citizenship are emerging. Echoing Yuval-Davis's (1999) 'multi-layered' citizenship, Ong's (1999) 'flexible' citizenship has been forwarded as a notion of flexible loyalties that transcend any particular nation-state. Joseph's (1999) 'nomadic' citizenship expands on Arjun Appadurai's (1990) topography of scapes (ethnoscape, technoscape, mediascape, etc.) to include that of 'citizenscape' as a means of theorizing identity for dislocated communities, refugees, and immigrant populations. Joseph's (1999) use of citizenscape enables the analysis of overlapping and multi-faceted narratives of identities. She reiterates the performative nature of citizenship as 'a ceaseless activity in which the fragments of various nations are scraped together into a makeshift home' (*Ibid.*). Bullen and Whitehead (2005, pp. 513–514) also argue for a 'sustainable', *fin de siècle* post-cosmopolitan citizenship that recognizes the importance of time and the world beyond the purely human.

In the global context, communities 'should be understood as relational spaces, composed of myriad networks of socio-ecological flows, stretching across various global and local scales' (*Ibid.*, p. 507). It is this politics of connectivity to other times and most particularly to other places, which, in the words of Desforges *et al.* (2005, p. 444) gives shape to 'some of the most important, and potentially liberating, new geographies of citizenship in the contemporary world... In this way, both local and extralocal interested and affected actors should be able to contribute to particular political programmes and visions of citizenship'. In this context, the relevance of physical public spaces increases as important platforms of citizenship that articulate the local, the national, and the global. Spaces of media and the Internet can alternatively compete with these public spaces, complement them, or reinforce their dominance. These concepts – multi-layered, flexible, nomadic, performative, sustainable citizenship, etc. – shed light on the multifaceted ideals with which city residents may identify. Individuals may find a way of reconciling their different loyalties with each other, but for many, this constitutes an ongoing challenge.

Democracy

Democracy is defined as a political system with majority rule and a separation of executive, judicial and legislative powers. The popular definition of a system 'by the people, for the people' has gradually prompted the expansion of democracy to mean decentralization, citizen participation, social justice, and respect of minorities (Tomas, 2004, p. 162). In Latin America, an important factor in the saliency of democratic and citizenship concerns has been the maturing of urbanization throughout the continent. As Roberts (2005, p.144) reports, 'by 2000, 75 percent of the Latin American population was urban, and most of the urban population lived in cities of over 100,000 people'. In addition, the pronounced urban primacy whereby one city, usually the capital, houses a large proportion of the national population compounds the importance of extraordinary events in public spaces. In this book, we take the stance advanced by Foweraker (2005) that the social mobilization for rights in public space holds the key to the improvement of democratic governance in Latin America.

This perspective differs from the view of democracy as the historical result of 'good lobalisa' in the form of the civic community (Putnam, 1993). This latter view takes 'civicness' as a functional prerequisite for democracy rather than exploring the popular agency that may achieve or improve it... An emphasis on social mobilization, in contrast, suggests that it is 'bad lobalisa' in the form of the fight for rights that can do most to improve the quality of democracy and deliver its substance to the citizenry at large... In the near future it is not likely to be a democracy made in the image of a perfectly civic society. But social mobilization may achieve the political conditions for piecemeal social development and greater efficacy in the rule of law. (Foweraker, 2005, p. 135)

'Bad behaviour', in Foweraker's ironic terms, is akin to the notions of insurgent citizenship and invented spaces of citizenship described above. These new expansive and performative types of citizenship beg a radical redefinition of democracy. In principle, deliberative democracy could offer a resolution to the performative paradox – the gaps between legal rights and actual practices of citizenship. In this approach, ethical principles emerge out of collective and fair deliberative processes in the public sphere in which arguments are constructed, discussed, and evaluated until the best prevail (Habermas, 1996). However, deliberative democracy has been criticized for not accounting for power imbalances and issues of exclusion and inequality (Young, 1996). 'Radical democracy', which redefines the liberal democratic principles of equality and liberty, may redress these shortcomings without diluting the differences and interests of diverse groups in the name of consensus. As Cohen and Fung explain,

In particular, radical-democratic ideas join two strands of democratic thought. First, with Rousseau, radical democrats are committed to broader participation in public decision-making... Second, radical democrats emphasize deliberation. Instead of a politics of power and interest, radical democrats lobal a more deliberative democracy in which citizens address public problems by reasoning together about how best to solve them. (Cohen and Fung, 2004, pp. 23–24).

In their poststructuralist critique of 'liberal essentialism', Ernesto Laclau and Chantal Mouffe (1985) criticized Rawls and other liberal theorists for essentializing identities and norms under the guise of rationality and neutrality (Mouffe, 1995). For Mouffe, no one should control the fundamental principles of society because that would define and silence the excluded. What democracy should provide instead, she claims, is a 'grammar of conduct' for people to abide by.

But more than the attainment of an ideal of radical democracy, what interests us here is in the *radicalization of democracy*. This entails different trajectories for each city and country in a context-specific search for a just city or nation. The collective imagining and mapping of such tailored trajectories in public space and all other venues in the public sphere, and the actual traversing of those paths, are what can ultimately help achieve the best conditions possible for full participation *and* deliberation.

What might those paths for the radicalization of democracy look like in each case studied in this book? This is an ongoing question that should be addressed through public participation and deliberation in each of those places. Tentatively, however, we may assume that a radicalization of democracy in Havana, for example, would expand the freedom of expression and the right to dissent while protecting the rights of Cubans to select and maintain their politico-economic system. In Caracas, it would entail respect for supporters and dissenters of the political regime, while facilitating progress in the transformation of a representative to a participatory democracy. In Rio de Janeiro, it would make all people accountable to the rule of law, while expanding human development opportunities for the disenfranchised. The same in Buenos Aires would clarify the crimes of the past, reconcile people with their present and with each other, and work for an inclusive and just future. In São Paulo, the process needs to respect the existing physical spaces of invited citizenship while trying to expand and appreciate opportunities for invented spaces of citizenship through design and policy-making. A radicalization of democracy in Santo Domingo would affirm respect for human rights and the rule of law through appropriations of space as Dominicans collectively move forward in overcoming the excesses and abuses of the past. In Santiago de Chile, it would facilitate the articulation of collective subjectivities to spaces that help Chileans engage in proactive agency to secure a future free of social and political repression.

If history is any indication, the people in Latin America will have to keep struggling for change to move in the direction of realizing these visions. In this venture, public space can be both a springboard for these mobilizations and an indicator of the sincere commitment to democracy on the part of those that create, maintain, regulate, and use these spaces. Meanwhile, we share Kohn's hope that 'a careful analysis of sites of resistance … might strengthen a conception of democracy that is useful today' (Kohn, 2003, p. 2).

Public Space

We recognize that the public has come to encompass the a-spatial world of the media, the Internet, and other trans-local conduits, but we aim to recover a focus on the physical space of plazas, streets, boulevards, parks, beaches, etc. We also continue a tradition of equating public and urban in our analyses of space. 'Stretching back to Greek antiquity onward, public space is almost by definition urban space, and in many current treatments of public space the urban remains the privileged scale of analysis and cities the privileged site' (Low and Smith, 2006, p. 3). There are several criteria in the making of 'publicness'. First, the public refers to that which is general, collective, and common. Second, public is that which is visible and ostensible. Lastly, public is that which is open and accessible to all (Rabotnikof, 2003). Public spaces facilitate encounters, and thus social learning.

Public spaces embody the tension between cultural diversity and social integration, and are crucial to the expression and resolution of complex socio-spatial transformations in cities around the globe. Discussions about public space try to address the need to strengthen both the sense of citizenship amidst the fragmentation of identities and the acknowledgment and celebration of plurality (Ramírez Kuri, 2003). The ideal of public spaces – open, accessible, inclusive, and capable of supporting respectful encounters of differences – makes them privileged sites in this quest (Makowski, 2003). As common ground for sociability and conflict, public spaces are terrain for the dialogical and dialectical practices of citizenship. The symbolic aspects of public spaces, a collective imaginary of memories, histories, and meanings, complements the physical characteristics of places (*Ibid*.). The places analysed in this book play a definite role in the construction of 'imagined communities' in the nations they belong to (Anderson, 1983). As an example, Alejandro Encinas, former Secretary of the Environment in Mexico City, asked those living in the vicinity of the Zócalo if they wanted the plaza landscaped. Though approved by voters, he faced protests from those claiming that this was not a decision for neighbours or even the city to make, because the space belonged to the whole nation (Tomas, 2004, p. 169). Citizenship and public space are tightly intertwined and to a great extent define each other, as 'both are the result of the interactions and struggles to generate and expand citizenship spaces' (Tamayo, 2004, p. 154). Not only a *mise-en-scène* for diversity and difference, public spaces are sites for the negotiation of values, rights, duties, and rules of sociability in a community.

Ideally, public space has to be multifunctional and capable of stimulating symbolic identification and cultural expression and integration (Borja, 2003*b*, p. 67). Regrettably, the recent growth of most Latin American cities has occurred without much expansion of public spaces. On the contrary, the prevalent trend has been to focus on transport infrastructure, shopping centres, and exclusive communities. The loss in quantity and quality of public space has affected the quality of life of city residents. Spaces abandoned by the middle- and high-income classes were colonized by the poor while others were renovated to serve the tourist and elite classes, as semi-private spaces were created within gated residential and business enclaves (Caldeira, 2000; Duhau, 2003). The resulting socio-spatial reorganization often results in 'the coexistence, without co-presence, of the poor with the middle- and high-income classes' (Duhau, 2003, p. 163). Most literature on public space with a focus on the United States and Europe also decries its privatization and commodification often linked with the expansion of the capitalist society. According to Low and Smith (2006, p. 4) in 'the Western world today, truly public space is the exception not the rule'. Nevertheless, it is important to acknowledge significant efforts that are being made in many Latin American cities to recover or create effective public space, signalling a 'renaissance of interest in public space in the current Latin American urbanism' (Borja, 2003*a*, p. 94).

How do extraordinary events transform public space?

The literature on the symbolic and spatial impacts of extraordinary events in public spaces is sparce. However, Foucault's notion of heterotopic space can provide a means to conceptualize public spaces as the evolving sites of extraordinary events. He contrasted the notion of utopias – idealized conceptions of society, inexistent in reality – with his inverse idea of heterotopias, or socially constructed counter-sites. Foucault (1997, p. 351) asserts 'There also exist … real and effective spaces … in which all the real arrangements … are at one and the same time represented, challenged and overturned'. Heterotopias give rise to new discourses about knowledge, power, subjectivity, and space. These alternate worlds fracture and entangle time and space and simultaneously reconstitute social relationships. Elspeth Probyn, building on Foucault, argues that '[h]eterotopia juxtaposes in one real place several different spaces, "several sites that are in themselves incompatible" or foreign to one another ... these are ... "places where many spaces converge and become entangled"' (cited in Guertin, nd, pp. 10, 11).

It is important to acknowledge, as Harvey (2000) points out, that alterity, or radical difference from the dominant society, by itself does not produce resistance or even critique of the *status quo*. We then refer here to the notion of heterotopia of resistance: 'a real counter-site that inverts and counters existing economic or social hierarchies. Its function is social transformation rather than escapism, containment, or denial' (Kohn, 2003, p. 91). It is in extraordinary events that many of the places studied here become heterotopias of resistance, or socially constructed counter-sites. In everyday life, but even more intensely and unexpectedly during extraordinary events, spaces are actively produced and reproduced to sustain or alter socio-economic and cultural givens. Cupers (2005, p. 734) asserts that it is particularly during extraordinary events when 'spatiality in the city is profoundly impure and hybrid, that this process unfolds. Space can thus be envisaged as a palimpsest of historical layers, some of which have disappeared while others remain active in constituting identities'. Cupers, who extends Careri's (2002) notion of the 'nomadic city', also proposes that

Understanding space and identity in terms of their continual change, may lead to the concept of a *nomadic geography*... The nomadic character of identity and space gives rise to an architecture of events, an urbanism of the situation... As such, the city's nomadic geography guarantees the presence of a strangeness that is possibly the most essential characteristic of the city (Cupers, 2005, p. 737, his emphasis).

Careri and Cupers see the nomadic city as a space open to progressive politics. 'Here new forms of lobalisa appear, new ways of dwelling, and new spaces of freedom' (Careri, 2002, p. 188). Cupers refers to London's East End as an example of a space in which the strange is allowed within the interstices of the familiarity of settlement. Yet we can think of the Latin American public spaces explored in this volume as such places. Cupers also critiques urban planning as a technology of domination, a position echoed by many scholars (see Flyvbjerg and Richardson,

2002; Holston, 1989; Peattie, 1987; Irazábal, 2004, 2005; Beard, 2002, 2003; Sandercook, 1998a; Miraftab, 2004). The call for the recognition of the fundamental linkages between citizenship, democracy, and public space made in this book is a call for more democratic and liberating planning practice.

The politics of and in public space

Around the world today, people are using public spaces to express their frustration, their dissent, and their hopes for alternative socio-political projects. Evidently, there are historical precedents of this phenomenon, especially in times of crises. As Low and Smith (2006, p. 16) argue, 'political movements are always about place and asserting the right, against the state, to mass in public space'. The cases analysed in this book detail significant evidence that '[t]he neoliberalism of public space is neither indomitable nor inevitable', and that 'whatever the deadening weight of heightened repression and control over public space, spontaneous and organized political response always carries within it the capability of remaking and retaking public space and the public sphere' (Ibid., p. 16). Since dissent, a cornerstone of Western liberal democracies, always threatens to exceed its bounds and become a threat, a challenge facing liberal states has not only been how to incorporate dissent, but also how to shape and control it: 'The politics of public space is thus a politics of location: where voices are silenced makes a huge difference as to which voices are heard. The politics of public space, therefore, can shape the nature of politics in public space' (Mitchell and Staeheli, 2005, p. 798). The over-policing of dissent, however, has in some cases led to the neutralizing of the promise of protest. But it also seems that the forms of 'soft' restraint, such as is embodied in the US protest permit system, overlaid on zoning, and other spatial tactics, are no longer as useful because dissenting publics are defying them (Ibid.). The chapters in this book reassert the vitality and vibrancy of public space politics in Latin America in a world that is experiencing a significant decrease in opportunities for expression in public space. Latin America is thus a promising site for a 'grand refusal' arresting the march of global neoliberalism and asserting locally concerted ways of living.

Linkages between public space and public sphere

The chapters in this book also present an opportunity to explore further the linkages between the public sphere and public space. Despite the rising interest in exploring their relations (Fraser, 1990; Mitchell, 2003), the literature that links public space and public sphere rarely takes a spatial angle. Low and Smith make a strong argument for the respatialization of the public:

Where the weakness of the public space literature perhaps lies in the practical means of translation from theories of political and cultural economy to the materiality of public space … the weakness of the public sphere literature may lie in the distance that it maintains from the places and spaces of publicness … Once recognized, that spatiality of the public sphere potentially transforms our understanding of the politics of the public. An understanding of public space is an imperative for understanding the public sphere. (Low and Smith, 2006, p. 6)

The original understanding of the concept of the public sphere has to be revamped if it is to be a useful notion for understanding the current transformations of the contemporary city. The concept originally referred to the emerging class of bourgeois capitalists in the eighteenth century, who formed a sphere of private people coming together as a public, claiming the public sphere from the public authorities which made possible, for the first time, rational political debate between private people in which everyone in principle was able to participate (Habermas, 1989). Although this early concept of the public sphere continues to influence present-day theory, it has received many criticisms (Fraser, 1999; Garnham, 1999; Young, 1996), including in Latin America (Avritzer, 2002). I highlight the particular criticism of the relations of identity and space (Gould, 1996). Cupers explains,

In its blinding myth of abstract space and rational identity, the concept of the liberal public sphere ultimately fails to understand the complexities of space and identity formation in the contemporary city. As such, it projects a fixed geography that falls short in perceiving how ideologies determine the spaces of public discussion and negotiation, and how identity is formed beyond rationality. (Cupers, 2005, p. 731)

In the liberal concept of the public sphere, with its focus on rational communication, differences such as class, gender, and race become irrelevant, while in reality 'difference becomes a fundamental aspect of the ways in which people interact and express identities in urban space' (*Ibid.*, p. 731). Cupers and others contest the abstract universality and political rationality of the original understanding of the public sphere, and suggest that in our times the public sphere can instead be understood 'as a play of uncertain identities in contested spaces'. Several chapters in this book explicitly engage the notion of the public sphere to debunk the assumptions of political rationality and abstract universality. See, in particular, Lima and Pallamin, Tamayo and Cruz-Guzmán, and Irazábal and Foley.

The Right to the City

Henri Lefebvre's seminal work on 'the right to the city' (1968, 1996) refers to the rights to inhabit and make use of the city and the right to have rights regardless of one's legal status. Among planners, activists, academics, and NGOs the phrase has become increasingly popular spurring the emergence of many new ideas (some of which come from Latin American, e.g., Buroni 1998; and Daniel 2001; but also Isin, 2000 and Soja, 2000; among others). In Purcell's words:

Lefebvre (1991) maintains that space is implicated in all elements of social life, the right to control the decisions that produce urban space implies the right to determine the full scope of everyday life in the city. Lefebvre's right to the city thus envisions a thoroughgoing democractization of urban politics. He insists that decision-making about urban space should be guided by the principle that use value should always trump exchange value – that above all other considerations urban space should be produced to meet the everyday needs of those who inhabit it. (Purcell, 2005, p. 200)

… Lefebvre interrogates and rethinks the decision-making structures that produce the city, and so introduces a much more radical democratization of the city … not just the right to speak in public space, but to decide the geography of public space; not just the right to be housed, but to decide the geography of affordable housing. (Ibid., p. 201)

For Harvey (2003, p. 940), the right to the city should not be 'merely a right of access to what the property speculators and state planners define, but an active right to make the city different, to shape it more in accord with our heart's desire, and to re-make ourselves thereby in a different image'. But despite these possibilities, Purcell argues that the 'right to the city' has not been sufficiently operationalized. There is still much work to be done in order to develop fully 'how visionary radical theory might articulate with everyday struggles against oppression' (Ibid., p. 201). But in invited and invented spaces of citizenship in Latin America, residents are pushing for the expansion of their rights to the city.

Implications for Planning and Policy Education, Research and Practice

We hope to contribute to a greater engagement with space within interdisciplinary work on citizenship, democracy, and the public sphere. We also aim to promote a reconstitution of urban planning and design thought and practice so as to aspire to 'just cities' that facilitate the unencumbered development of full citizenship for all residents. As Miraftab (2004, p. 212) suggests, these stories of ordinary and extraordinary events in urban public spaces underline 'the significance of both invited and invented spaces of citizen participation in the formation of inclusive cities and citizenship'. The ideas discussed in this chapter – among them, nomadic geography, insurgent urbanism, and invented spaces of citizenship – can help us envision a progressive politics that translates into planning theory, education, and practice grounded in a sophisticated understanding of citizenship and a challenge to neoliberal urbanism (Miraftab, 2004, p. 202). This should lead planners and policy-makers to reassess their roles and to acknowledge and encourage the kind of citizenship practices that Holston and Appadurai (1999, p. 20) describe being able to create 'new kinds of citizenship, new sources of laws, and new participation in decisions that bind'.

In effect, the recent reinterpretations of the notion of citizenship covered in this chapter move planning theory beyond participatory planning to insurgent planning (Miraftab, 2004; Holston, 1995, 1998; Sandercock, 1998*a*, 1998*b*; Friedmann, 2002). This is a significant restructuring of the realm of planners' inquiry and commitment: 'If modernist planning relies on and builds up the state, then its necessary counter agent is a mode of planning that addresses the formations of insurgent citizenship' (Holston, 1998, p. 47). Also in Miraftab's (2004, p. 211) words: 'for an emerging wave of planners who take into account an expanded realm of citizenship construction, the sources of information and guidance for planning practices are the everyday spaces of citizenship', and we would also add, the extraordinary spaces of citizenship. Scholars advocate an epistemological shift within planning theory and education as a move to 'engage, in practice, the problematic nature of belonging to society' (Holston, 1999*a*, p. 173), or as Miraftab (2004, p. 211) claims, '[a] planning practice that relies not merely on the high commands of the state but on situated practices of citizens'. Urban designers, planners, and politicians have yet to confront these shifting socio-spatialities. For Sandercock (1998*b*), such 'radical planning' does not necessarily begin with grand acts, but instead with smaller actions that she calls 'a thousand tiny empowerments'. Significantly, some of the events discussed in this book simultaneously constitute grand acts *and* tiny empowerments. A next step in researching extraordinary events in public spaces would be to assess the conditions under which they result in an insurgent citizenship culminating in a better quality of life and an urbanism more responsive to the needs of city.

The explorations in this book are timely as we are witnessing a rebirth of interest in public space, and the reformulation of citizenship and democracy in Latin America as these countries resist neoliberal dictates, leaning left at a time when the United States is leaning right. In this world of interconnectedness, these polarizations are not isolated events, but their ultimate implications remain to be seen. The 'grand refusal' against neoliberalism in Latin America may prove to be short lived, but if it lasts, and while it lasts, it may bring critical restructuring not only to cities in the South but also to the system of global order. The seeds of a new world, or at least an alternative world order, may very well be in the making.

References

Amin, A. (2004) Regions unbound: towards a new politics of place. *Geografiska Annaler*, **86**(B), pp. 33–44.

Anderson, Benedict (1983) *Imagined Communities: Reflections on the Origin and Spread of Nationalism.* London: Verso.

Appadurai, Arjun (1990) Disjuncture and difference in the global cultural economy. *Public Culture,* **2**(2), pp. 1–24.

Avritzer, Leonardo (2002) *Democracy and the Public Space in Latin America.* Princeton: Princeton University Press.

Beard, Victoria (2002) Covert planning for social transformation in Indonesia. *Journal of Planning Education and Research,* **22**(1), pp. 15–25.

Beard, Victoria (2003) Learning radical planning: the power of collective action. *Planning Theory,* **2**(1), pp. 13–35.

Borja, Jordi (2003*a*) Ciudad y planificación: La urbanística para las ciudades de América Latina. in Balbo, Marcelo, Jordán, Ricardo and Simioni, Daniela (eds.) *La Ciudad Inclusiva.* Santiago de Chile: Naciones Unidas.

Borja, Jordi (2003*b*) La ciudad es el espacio público, in Kuri, Ramírez (ed.) *Espacio Público y Reconstrucción Ciudadana.* Mexico: Flacso and Miguel Angel Porrúa Grupo Editorial.

Borja, Jordi and Muxí, Xaida (2002) *Espacio Público, Ciudad, y Ciudadanía.* Madrid: Alianza Editorial.

Boudreau, Julie-Anne (2000) *The Megacity Saga: Democracy and Citizenship in this Global Age.* Montreal: Black Rose Books.

Bullen, Anna and Whitehead, Mark (2005) Negotiating the networks of space, time and substance: a geographical perspective on the sustainable citizen. *Citizenship Studies,* **9**(5), pp. 499–516.

Buroni, T. (1998) A Case for the Right to Habitat. Paper presented at Seminar on Urban Poverty, Rio de Janeiro.

Caldeira, Teresa (2000) *City of Walls: Crime, Segregation, and Citizenship in São Paulo.* Berkeley, CA: University of California Press.

Cantú Chapa, Rubén (2005) *Globalización y Centro Histórico: Ciudad de México, Medio Ambiente Sociourbano.* Mexico: Plaza y Valdés.

Capron, Guénola and Monnet, Jérôme (2003) Una retórica progresista para un urbanismo conservador: La protección de los centros históricos en América Latina, in Ramírez Kuri, Patricia (ed.) *Espacio Público y Reconstrucción de Ciudadanía.* Mexico: Flacso and Miguel Angel Porrúa Grupo Editorial.

Careri, F. (2002) *Walkscapes: Walking as an Aesthetic Practice.* Barcelona: Gustavo Gili.

Cohen, Joshua, and Fung, Archon (2004) Radical democracy. *Swiss Journal of Political Science,* **10**(4), pp. 23–32.

Crang, Mike, Crang, Phil and May, Jon (1999) *Virtual Geographies: Bodies, Space, and Relations.* London: Routledge.

Cupers, Kenny (2005) Towards a nomadic geography: rethinking space and identity for the potentials of progressive politics in the contemporary city. *International Journal of Urban and Regional Research,* **29** (4), pp. 729–739.

Daniel, Celso (2001) Participatory urban governance: the experience of Santo Andre. *United Nations Chronicle Online,* **38**(1).

Davis, Mike (1990) *City of Quartz: Excavating the Future in Los Angeles.* New York: Verso.

Delanty, Gerard (2002) Two conceptions of cultural citizenship: a review of recent literature on culture and citizenship. *The Global Review of Ethnopolitics,* **1**(3), pp. 60–66.

Desforges, Luke, Jones, Rhys and Woods, Mike (2005) New Geographies of citizenship. *Citizenship Studies,* **9**(5), pp. 439–451.

Douglass, Mike, and Friedmann, John (eds.) (1998) *Cities and Citizens.* New York: John Wiley.

Duhau, Emilio (2003) Las megaciudades en el siglo XXI: De la modernidad inconclusa a la crisis del espacio público, in Ramírez Kuri, Patricia (ed.) *Espacio Público y Reconstrucción de Ciudadanía.* Mexico, DF: Flacso and Miguel Angel Porrúa Grupo Editorial.

Fainstein, Susan (2000) New directions in planning theory. *Urban Affairs Review,* **35**(4), pp. 451–478.

Flacks, Richard (1988) *Making History vs. Making Life.* New York, NY: Columbia University Press.

Flyvbjerg, Bent and Richardson, Tim (2002) Planning and Foucault: in search of the dark side of planning theory, in Allmendinger, Philip and Tewdwr-Jones, Mark (eds.) *Planning Futures: New Directions for Planning Theory*. London: Routledge. pp. 44–62.

Foucault, Michel (1997) Of other spaces: utopias and heterotopias, Leach, Neil (ed.) *Rethinking Architecture: A Reader in Cultural Theory*. London: Routledge.

Foweraker, Joe (2005) Toward a political sociology of social mobilization in Latin America, in Wood, Charles H. and Roberts, Bryan R. (eds.) *Rethinking Development in Latin America*. University Park, PA: Pennsylvania State University Press.

Fraser, Nancy (1990) Rethinking the public sphere: A contribution to actually existing democracy. *Social Text*, Nos. 25/26, pp. 56–79.

Fraser, Nancy (1999). Rethinking the public sphere: A contribution to the critique of actually existing democracy, in Calhoun, Craig (ed.) *Habermas and the Public Sphere*. Cambridge, MA: MIT Press.

Fraser, Nancy, and Gordon, Linda (1994) Civil citizenship against social citizenship, in Steenbergen, B. Van (ed.) *The Condition of Citizenship*. London: Sage, pp. 90–107.

Friedmann, John (1998) The new political economy of planning: The rise of civil society, in Douglass, M. and Friedmann, J. (eds.) *Cities for Citizens: Planning and the Rise of Civil Society in a Global Age*. New York, NY: Wiley.

Friedmann, John (2002) *The Prospect of Cities*. Minneapolis, MN: University of Minnesota Press.

Friedmann, J. and Douglass, M. (1998) Editors' introduction, in Douglass, M. and Friedmann, J. (eds.) *Cities for Citizens: Planning and the Rise of Civil Society in a Global Age*. New York, NY: Wiley.

García Canclini, Néstor (1995) *Consumidores y Ciudadanos*. México: Grijalbo.

Garnham, Nicholas (1999) The media and the public sphere, in Calhoun, Craig (ed.) *Habermas and the Public Sphere*. Cambridge, MA: MIT Press.

Gaventa, John (2002) Exploring citizenship, participation and accountability. *IDS Bulletin*, **33**(2), pp. 1–11.

Gould, C.C. (1996) Diversity and democracy: representing differences, in Benhabib, S. (ed.) *Democracy and Difference: Contesting the Boundaries of the Political*. Princeton, NJ: Princeton University Press.

Graham, Stephen and Marvin, Simon (2001) *Splintering Urbanism: Networked Infrastructures, Technological Mobilities, and the Urban Condition*. London: Routledge.

Gregory, Derek (1994) *Geographical Imagination*. Oxford: Blackwell.

Guertin, Carolyn G. (nd) *Quantum Feminist Mnemotechnics: The Archival Text, Digital Narrative and the Limits of Memory*. http://www.mcluhan.utoronto.ca/academy/carolynguertin/4i.html.

Habermas, Jürgen (1981) *The Theory of Communicative Action*. Boston, MA: Beacon Press.

Habermas, Jürgen (1989) *The Structural Transformation of the Public Sphere: An Inquiry into a Category of Bourgeois Society*. Cambridge: Polity Press.

Habermas, Jürgen (1996) Three normative models of democracy, in Benhabib, Seyla (ed.) *Democracy and Difference: Contesting the Boundaries of the Political*. Princeton, NJ: Princeton University Press.

Harvey, David (2000) *Spaces of Hope*. Berkeley, CA: University of California Press.

Harvey, David (2003) Debates and developments: the right to the city. *International Journal of Urban and Regional Research*, **27**(4), pp. 939–941.

Holston, James (1989) *The Modernist City: An Anthropological Critique of Brasília*. Chicago, IL: University of Chicago Press

Holston, James (1995) Spaces of insurgent citizenship. *Planning Theory*, **13**, pp. 35–52.

Holston, James (1998) Spaces of insurgent citizenship, in Sandercock, Leonie (ed.) *Making the Invisible Visible: A Multicultural Planning History*. Berkeley, CA: University of California Press.

Holston, James (ed.) (1999*a*) *Cities and Citizenship*. Durham, NC: Duke University Press.

Holston, James (1999*b*) Spaces of insurgent citizenship, in Holston, James (ed.) *Cities and Citizenship*. Durham, NC: Duke University Press.

Holston, James and Appadurai, Arjun (1999) Introduction, in Holston, James (ed.) *Cities and Citizenship*. Durham, NC: Duke University Press.

Holston, James and Appadurai, Arjun (1996) Cities and citizenship. *Public Culture*, **8**, pp.187–204.

Irazábal, Clara (2004) A planned city comes of age: rethinking Ciudad Guayana today. *Journal of Latin American Geography*, **3**(1), pp. 22–51.

Irazábal, Clara (2005) *City Making and Urban Governance in the Americas: Curitiba and Portland*. London: Ashgate.

Isin, Engin F. (1999) Cities and citizenship. *Citizenship Studies*, **3**(2), pp. 165–172.

Isin, Engin F. (ed.) (2000) *Democracy, Citizenship, and the Global City*. London: Routledge.

Isin, Engin F. (2002*a*) *Being Political: Genealogies of Citizenship*. Minneapolis, MN: University of Minnesota Press.

Isin, Engin F. (2002*b*) City, democracy and citizenship: historical images, contemporary practices, in Isin, E.F. and Turner, B.S. (eds.) *Handbook of Citizenship Studies*. London: Sage.

Joseph, May (1999) *Nomadic Identities: The Performance of Citizenship*. Minneapolis, MN: University of Minnesota Press.

Kabeer, Naila (2002) Citizenship, affiliation, and exclusion: Perspective from the south. *IDS Bulletin*, **33**(2), pp. 12–23.

Kearns, A. (1995) Active citizenship and local governance: political and geographical dimensions. *Political Geography*, **14**, pp. 155–175.

Kohn, Margaret (2003) *Radical Space: Building the House of the People*. Ithaca, NY: Cornell University Press.

Kohn, Margaret (2004) *Brave New Neighborhoods: The Privatization of Public Space*. New York, NY: Routledge.

Laclau, Ernesto and Mouffe, Chantal (1985) *Hegemonist and Socialist Strategy: Towards a Radical Democratic Politics*. London: Verso.

Lechner, Norbert (2000) Nuevas ciudadanías. *Revista de Estudios Sociales*, No 5.

Lee, Benjamin (1998) Peoples and publics. *Public Culture*, **10**(2), pp. 371–394.

Lefebvre, Henri (1968) *Le droit á la ville*. Paris: Anthopos.

Lefebvre, Henri (1991) *The Production of Space*. Oxford: Blackwell.

Lefebvre, Henri (1996) *Writings on Cities* (trans. Kofman, E. and Lebas, E.). Oxford: Blackwell.

Low, Setha (2000) *On the Plaza: The Politics of Public Space and Culture*. Austin, TX: University of Texas Press.

Low, Setha and Smith, Neil (eds.) (2006) *The Politics of Public Space*. New York, NY: Routledge.

MacLeod, Gordon (nd) Citizenship in the 'post-justice' city, in Geography of Leisure and Tourism Research Group, *Geographies of Citizenship*. Royal Geographical Society (with the Institute of British Geographers). http://www.exeter.ac.uk/geography/tourism/gltrg/Events/London/citizenship.html

Makowski, Sara (2003) Alteridad, exclusion, y ciudadanía: Notas para una reescritura del espacio público, in Ramírez Kuri, Patricia (ed.) *Espacio Público y Reconstrucción de Ciudadanía*. Mexico: Flacso and Miguel Angel Porrúa Grupo Editorial.

Marshal, Thomas Humphrey (1964) *Class, Citizenship and Social Development*. Chicago, IL: Chicago University Press.

Massey, D. (2004) Geographies of responsibility. *Geografiska Annaler*, **86**(B), pp. 5–18.

McBride, Keally (2005) Book review of *Brave New Neighborhoods: The Privatization of Public Space* (Kohn, Margaret (2004) New York, NY: Routledge). *International Journal of Urban and Regional Research*, **29**(4), pp. 997–1009.

McDonnell, Patrick (2005) Populist Indian leads in Bolivian polls. *Los Angeles Times*, 18 December, p. A3.

Miraftab, Faranak (2004) Invented and invited spaces of participation: neoliberal citizenship and feminists' expanded notion of politics. *Journal of Transnational Women's and Gender Studies*, **1**(1). http://web.cortland.edu/wagadu/vol1-1toc.html (accessed December 2005).

Miraftab, Faranak and Wills, Shana (2005) Insurgency and spaces of active citizenship: the story of Western Cape anti-eviction campaign in South Africa. *Journal of Planning Education and Research*, **25**(2), pp. 200–217.

Mitchell, Don (2001) Postmodern geographical praxis? The postmodern impulse and the war against the homeless in the post-justice city, in Minca, Claudio (ed.) *Postmodern Geography: Theory and Praxis*. Oxford: Blackwell, pp. 57–92.

Mitchell, Don (2003) *The Right to the City: Social Justice and the Fight for Public Space*. New York, NY: Guilford Press.

Mitchell, Don, and Staeheli, Lynn A. (2005) Permitting protest: parsing the fine geography of dissent in America. *International Journal of Urban and Regional Research*, **29**(4), pp. 796–813.

Mouffe, Chantal (1995) Feminism, citizenship, and radical democratic politics, in Nicholson, L. and Seidman, S. (eds.) *Social Postmodernism*. Cambridge: Cambridge University Press.

Muxí, Zaida (2004) *La arquitectura de la ciudad global*. Barcelona: Editorial Gustavo Gili.

O'Donnell, Guillermo (1999) *Counterpoints: Selected Essays on Authoritarianism and Democratization*. Notre Dame, IN: University of Notre Dame Press.

Ong, A. (1999) *Flexible Citizenship: The Cultural Logics of Transnationality*. Durham, NC: Duke University Press.

Peattie, Lisa (1987) *Planning: Rethinking Ciudad Guayana*. Ann Arbor, MI: University of Michigan Press.

Purcell, Mark (2005) Review of *The Right to the City: Social Justice and the Fight for Public Space* by Don Mitchell. *Antipode*, **37**(1), pp. 199–202.

Putnam, Robert (1993) *Making Democracy Work: Civic Traditions in Modern Italy*. Princeton, NJ: Princeton University Press.

Rabotnikof, Nora (2003) Introducción: Pensar lo público desde la ciudad, in Ramírez Kuri, Patricia (ed.) *Espacio Público y Reconstrucción de Ciudadanía*. Mexico: Flacso and Miguel Angel Porrúa Grupo Editorial.

Ramírez Kuri, Patricia (2003) *Espacio Público y Reconstrucción de Ciudadanía*. Mexico: Flacso and Miguel Angel Porrúa Grupo Editorial.

Roberts, Bryan (2005) Citizenship, rights, and social policy, in Wood, Charles H. and Roberts, Bryan R. (eds.) *Rethinking Development in Latin America*. University Park, PA: Pennsylvania State University Press.

Rose, Nikolas (2000) Governing cities, governing citizens, in Isin, Engin F. (ed.) *Democracy, Citizenship and the Global City*. London: Routledge.

Rosenthal, Anton (2000) Spectacle, fear, and protest: a guide to the history of urban public space in Latin America. *Social Science History*, **24**(1), pp. 33–73.

Roy, Ananya (2001) A 'public' muse: on planning convictions and feminist contentions. *Journal of Planning Education and Research*, **21**, pp. 109–126.

Roy, Ananya (2003) *City Requiem, Calcutta: Gender and the Politics of Poverty*. Minneapolis, MN: University of Minnesota Press.

Ruggiero, Vincenzo (2005) Dichotomies and contemporary social movements. *City*, **9**(3), pp. 297–306.

Sandercock, Leonie (1998*a*) Framing insurgent historiographies for planning, in Sandercock, Leonie (ed.) *Making the Invisible Visible: A Multicultural Planning History*. Berkeley, CA: University of California Press.

Sandercock, Leonie (1998*b*) *Towards Cosmopolis*. New York, NY: Wiley.

Sassen, Saskia (1996) Whose city is it? *Traditional Dwellings and Settlements Review*, **8**(1), p. 11.

Scarpaci, Joseph (2005) *Plazas and Barrios: Heritage Tourism and Globalization in the Latin American Centro Histórico*. Tucson, AZ: University of Arizona Press.

Smith, Neil (1996) *The New Urban Frontier: Gentrification and the Revanchist City*. London: Routledge.

Soja, Edward (2000) *Postmetropolis: Critical Studies of Cities and Regions*. Oxford: Blackwell.

Sorkin, Michael (ed.) (1992) *Variations on a Theme Park: The New American City and the End of Public Space*. New York, NY: Hill and Wang.

Tamayo, Sergio (2004) Espacios ciudadanos, in Rodríguez, Ariel and Tamayo, Sergio (eds.) *Los últimos cien años, los próximos cien*. México DF: Universidad Autónoma Metropolitana.

Tomas, François (2004) Espacios públicos que convierten la metrópolis de nuevo en ciudad, in Rodríguez, Ariel and Tamayo, Sergio (eds.) *Los últimos cien años, los próximos cien*. Mexico: Universidad Autónoma Metropolitana.

Touraine, Alain, Dubet, François, *et al.* (1996) *Le grand refus: Réflexions sur la grève de décembre 1995*. Paris: Fayard.

United Nations Human Settlements Programme (2005) *The State of the World's Cities 2004/2005. Globalization and Urban Culture*. London: UN-Habitat and Earthscan.

Wark, McKenzie (1994) *Virtual Geography: Living With Global Media Events*. Bloomington, IN: Indiana University Press.

Wekerle, Gerda (2000) Women's rights to the city: gendered spaces of pluralistic citizenship, in Isin, Engin F. (ed.) *Democracy, Citizenship and the Global City*. London: Routledge.

Winocur, Rosalía (2003) La invención mediática de la ciudadanía, in Ramírez Kuri, Patricia (ed.) *Espacio Público y Reconstrucción de Ciudadanía*. Mexico: Flacso and Miguel Angel Porrúa Grupo Editorial.

Young, Iris Marion (1990) *Justice and the Politics of Difference*. Princeton, NJ: Princeton University Press.

Young, Iris Marion (1996) Communication and the other: beyond deliberative democracy, in Benhabib, Seyla (ed.) *Democracy and Difference: Contesting the Boundaries of the Political*. Princeton: Princeton University Press

Yuval-Davis, Nira (1999) The multi-layered citizen: citizenship at the age of 'glocalization'. *International Feminist Journal of Politics*, **1**, pp. 119–136.

Zukin, Sharon (1991) *Landscapes of Power: From Detroit to Disney World*. Berkeley, CA: University of California Press.

Chapter 2

Political Appropriation of Public Space: Extraordinary Events in the Zócalo of Mexico City

Sergio Tamayo and Xóchitl Cruz-Guzmán

Introduction

Mexico City, historically the most important social and political centre in Mexico, was the scene of two extraordinary events in 2000 and 2001 with impacts for the country's democratic life. The first was the presidential election campaign; the second, the public mobilization of the Zapatista Army for National Liberation (EZLN). The main plaza of the capital, the Zócalo, was transformed for these two great occasions. Citizens without access to formal institutional channels used the public space of the Zócalo to express themselves. Groups and social classes who were at odds with one another took to the streets for protest and participation. These appropriations of public space deepened the connections between space, social practices, and political conflict.

One significant characteristic of the Zócalo is that it provides material representation of the three great societal powers since the time of the Spanish colony: the Catholic church, in the form of the metropolitan cathedral; the state, represented by the national palace and the palace of the local government; and the entrepreneurial elite, symbolized by the commercial centres and great hotels. It is also significant that the Zócalo was built on the ruins of Tenochtitlan, the capital of the Aztec Empire and destroyed by the Spanish in 1521. Remains of the city's Templo Mayor (Great Temple) were discovered in 1978 during the construction of a metro line. The Zócalo also represents contemporary Mexico. It is a centre of national identity (Wildner, 1998, 2005).

Mexico City is among the largest cities of the world. It is a cosmopolitan city, home to the federal powers, and the main node of international exchange in the country. With millions of domestic migrants, the population of the metropolitan area reached 18.7 million in 2003. Mexico City is also a marker both of modern industrialism and of some problematic conditions of post-modern urbanism – e.g., spatial fragmentation and social polarization. The city has been the scene of strong confrontation between groups with distinct social and political aims. Each of these projects has provided sustenance for different utopias and visions born in the city itself or throughout the nation. The Zócalo has become a symbolic space where major national unions, regional cattle-ranching organizations, state political movements, insurgent indigenous movements, local urban and civic organizations, and religious groups have literally found 'common ground'.

Here we analyse two extraordinary cases of political contention: an electoral struggle and a Zapatista mobilization. Each has represented a distinct political and social appropriation of ordinary space. We critique every-day-life culturalist visions of public space, which in our view tend to ignore the importance of politics in the production of urban space. In the same vein, we critique those visions which overlook concrete space where public debate is crystallized. Physical place cannot be lost in political thought. Such approaches run the risk of fragmenting the appreciation and interpretation of reality. For a more comprehensive analysis, we emphasize the role of actors, values, mass media, and physical space. In the first part of this chapter, we reflect on the concept of the public sphere, its semantic correspondence with the notion of space and the constitution of this sphere, to understand public space as a product of forms of domination and social conflict. Secondly, we describe the political appropriation of public space at the close of the electoral campaign of the Christian Democratic Party for National Action (PAN) in 2000. Thirdly, we emphasize the production of political space through a discussion of the Zapatista march from Chiapas to Mexico City that took place shortly after, in 2001. Finally, we analyse the conditions of the production of public space in Mexico City.

The Public Sphere

The political appropriations of the Zócalo by the citizenry, political parties, social organizations, and civil society clearly illustrate the connection that exists between public space, citizen practices and experiences, and the confrontation and struggles that take place between groups from different social classes. Other studies in Mexico and Latin America in the last 10 years have emphasized the role of citizen participation in the formation of public space.[1] Laura Gingold (2000) synthesizes these studies in Latin America. Public opinion and the role of mass media provide a basis for her analysis, which builds on the liberal conceptualizations of Hanna Arendt, John Rawls, and Jürgen Habermas. Liberal definitions of the public sphere, assume the existence of harmonious dialogues, and suppose a public sphere beyond history (see figure 2.1). For liberals, it was important to struggle against the authoritarianism of the state. Thus, public sphere's principal values became the use of reason and the exercise of autonomy. In this way the public would have to constitute itself in a realm between state power and the private interests of individuals.

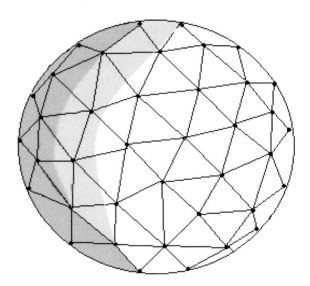

Figure 2.1. Liberal public sphere, ordered with communicational equidistance and consensual flows.

The public sphere is not only tied to values such as freedom and the individual capacity for reasoning and autonomy, but has two other attributes, as Hanna Arendt posits. It is a link to political power and a context for action. Politics contributes to the formation of the public sphere to the extent that individuals can freely devote themselves to the interests of the general public. Arendt's idea of action is based on activities that directly put human beings in relation with one other. Such activities correspond to the human desires for plurality and social interaction (Arendt, 1961; Collin, 2001).[2] For his part, Habermas's position can be inferred in Arendt's axioms concerning action. The public sphere is a discursive space for citizen debate, deliberation, agreement, and negotiation – a space of communicative action. Habermas argues that the space formed in the interstices of the state and civil society has an institutional character and is crucial to forming public opinion.[3]

This perspective that defines the public sphere as a homogenous body formed between free and equal individuals has been subjected to strong criticism. The most outstanding critique of Habermas rejects his idealization of the bourgeois public sphere and the ideal of discursive action that takes place in a rational manner and between equals. The appropriation of the Zócalo for electoral and socio-cultural politics is illustrative of a public sphere: where space for discussion is limited, mass media has an increased role in shaping of individual reasoning, and there is an unequal possession of symbolic resources for public argumentation. This is the crux of the problem in Habermas's theory of discursive action, i.e., it is anchored in an ideal of communication that ignores domination and struggle for power (Voirol, 2003). Two alternative ideas are introduced through this discussion of the public sphere: one opposes this liberal tradition, while the other speaks of a diversity of spaces – multiple public spheres – with diverse degrees of tolerance, dialogue, and consensus. This last position demands that we think of proletarian spheres, which could be established in contrast to a bourgeois sphere, and consequently create openings for traditionally excluded groups.[4] The problem of this vision is its ambiguity.

We nevertheless consider Habermas's concept of 'internal colonization' and the uncoupling of 'system and lifeworld' to be revealing of the processes of domination of the state over civil society. A revisionist Habermas (1993) emphasizes that today's public sphere is made up of political parties and civic associations. This represents a strategic perspective for regenerating public communication. As can be witnessed in various cases in Mexico City, the public sphere is more a product of social organizations than of individuals. However, the masses continue to struggle for the right to space and against the complexity of the state bureaucracies. Thus, we emphasize Axel Honneth's important critique of Habermas in which he stresses the role of conflict – including physical violence, exclusion, and public humiliation – in the formation of the public sphere and the collective subject. The experience of Mexico City shows that the dynamics of public space formation are a consequence of social struggle (Smelser, 1995).[5] They may go beyond tensions between the system and lifeworld, Honneth says (1996; 2000), and become a systematic violation of the rights of social recognition.[6]

Thus the public sphere in Mexico is a space of plurality, not of homogeneity. It is one which facilitates dialogue, but also channels and reproduces power. In a hermeneutic perspective, it is a result of struggles and negotiations between stratified traditions and languages (Alejandro, 1993). For example, electoral campaigns in Mexico are conceived within the tradition of representative democracy. But the mobilization of society and citizen groups and ensuing political confrontation in 2000 led to true political battles which undid the hegemony of the previous political regime of the Institutional Revolutionary Party (PRI) that had lasted 70 years. Similarly, in the case of the Zapatista march, indigenous Mexicans were able not only to impose national and international visibility, but also communicate their demands with a certain level of effectiveness.

These appropriations of space were political as much as spatial strategic positionings within the national public sphere. The public sphere, in this view, is a combination and juxtaposition of places constructed and sustained by different political projects, and a combination and juxtaposition of languages and power differentials, akin to a Foucauldian heterotopia (Foucault, 1994). In our case studies, the space of the Zócalo was the scene of performances where different political and social actors faced each other, as the following ethnographic account will show.

The Electoral Struggle

For most of the twentieth century, the PRI had monopolized Mexican politics. Nevertheless, at the end of the 1930s, the National Action Party (PAN) would arise from the right to resist what they identified as the totalitarianism of the state. It was not until the end of the 1980s, during the transition towards a neoliberal economy, that the PAN was strengthened by the participation of the business sector and segments of the middle-class (Tamayo, 1999). Another party with Social-Democratic leanings, the Party of Democratic Revolution (PRD), also arose in 1989. The PRD sheltered PRI dissidents and attracted various left-leaning groups such as the communists, Trotskyites, Maoists, Guevarists, and has gradually become the third political force of the country (*Ibid.*).

On 2 July 2000, Mexicans surprised the world by ending the 71 year PRI rule. It was an election that ended one of the most direct inheritances of the Mexican Revolution of 1910. Nevertheless, the electoral campaign 2000 favoured an ideologically conservative project personified by Vicente Fox, the PAN candidate. The mass media hailed the 'exemplary elections for Latin America', but the changes since then have been much more complex and paradoxical. The presidential campaigns of 2000 progressed in an atmosphere of great confrontation. At the national level the topics were the economy and finances, the conflict within the university student movement at the National Autonomous University of Mexico, political party struggles, defence of the laicism of the state, conflict in Chiapas with the *Ejército Zapatista de Liberación Nacional* (EZLN), and charges of electoral crime. Electoral activity became a declaration of war, and political analysts resorted to the use of opinion surveys, only recently introduced to Mexico, as a means of manipulation (Covarrubias, 2000).[7] Verbal violence was propagated electronically. The parties preferred mass media to the expensive work of facing voters directly. They paid for publicity time on the main Mexican television networks, Televisa and TV-Azteca, as well as on radio stations. The PAN and the PRI concentrated their efforts on the diffusion of the image of their candidates and to sloganeering, while the PRD seemed to concentrate on the mobilization of masses, prioritizing the concentrations in plazas with the aim of 'reorganization of the social fabric'.

In such an atmosphere of media aggressiveness, direct contact with the electorate at mass rallies, visits to people's homes, and meetings with social organizations were also important. For this reason, plazas during mass rallies became perceived as 'political thermometers'. At times, the practical value of the public concentrations was overestimated. According to Marcelino Parelló, an ex-student leader of the 1968 movement, political rallies in plazas 'do not win elections'; at best, 'they provide 100 thousand votes, [but] the million voters are in their houses, watching TV'. Public concentrations and events, however, have their importance, because they fortify political networks. The aim is not to predict electoral results with these events, but to understand the dynamics of urban and citizen cultures that are expressed and activated in public physical space. This spatial visibility allows for connections with the electorate and establishment of relations with social organizations. Ultimately, events in public space also make for powerful TV footage.

The Public Plaza

The public concentrations not only revealed the existence of three political alternatives for the city and the nation (i.e., the PRI, the PAN, and the PRD), but also how these projects were expressed culturally and ideologically. The Zócalo was the scene of the closing ceremonies of the campaigns of the three major parties. The citizen participation was unparalleled given the large numbers of people at the three rallies. Participation in the closing ceremony of the PAN electoral campaign was particularly notable given its duration, its sentimentality, and spontaneity.[8] Multitudes congregated around many of the landmarks and postmodern buildings throughout the route towards the main plaza.[9] Avenida Juárez, a historic tree-lined avenue, became an enormous parking lot for buses and cars whose passengers waited their turn to enter the march.

The architecture of the central plaza added to this sensational and exciting event. People crowded around each other and jammed themselves between magnificent buildings: the National Palace, the Metropolitan Cathedral, and the Palace of Government of the Federal District. The Majestic Hotel and the Holiday Inn around the Zócalo were filled; people crowded each window and balcony. Blue and white flags, PAN's symbolic colours, adorned the surrounding terraces. The concentration was a ritualized political act. However, it also became a multimedia spectacle which was at times devoid of any political essence. In the Zócalo the distribution of the participants and the material objects reflected the type of act that PAN had planned (see figure 2.2). In the background was a gigantic screen that projected the image of the speakers, particularly that of Vicente Fox. The stage was semi-enclosed and semi-covered. Its back faced the government offices of the Federal District, which was occupied by PRD, leftist opponents. Towards the right was the National Palace, from where Fox would govern, given political victory.

Those who looked up to the sky saw giant screens, aerostatic globes and blimps, as well as the helicopters of the presidential staff. But people focused on one point, the stage. The crowd was anxious to see its leader. The energy, emotion, and loudness were intensifying. Blond haired families and dark-skinned families, European descendents and the racially mixed were united under political conservatism. The PAN event was a citizen celebration devoid of trade unions and social organizations. It was a festive event where professionals and political specialists sought to affect public opinion.[10] In spite of the participatory element of the event which tried to break with the traditional corporatism of PAN, social class differences became readily visible. People who stood under the stage were

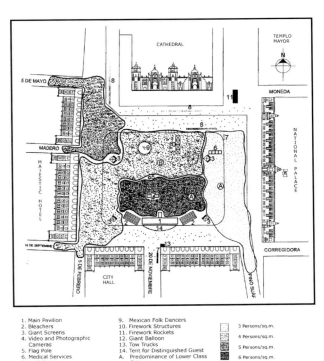

Figure 2.2. Physical and social appropriation of the space. Closing of campaign of Alliance for Change (PAN), 24 June 2000. (*Source*: Tamayo, 2002)

1. Main Pavilion
2. Bleachers
3. Giant Screens
4. Video and Photographic Cameras
5. Flag Pole
6. Medical Services
7. Ambulances
8. Street Vendors
9. Mexican Folk Dancers
10. Firework Structures
11. Firework Rockets
12. Giant Balloon
13. Tow Trucks
14. Tent for Distinguished Guest
A. Predominance of Lower Class Citizens
B. Predominance of Middle Class Citizens

3 Persons/sq.m.
4 Persons/sq.m.
5 Persons/sq.m.
6 Persons/sq.m.

Digitalización CAD: Consuelo Córdoba F

reflective of a mixture of social classes, mainly the popular and lower middle-class who were attracted by Fox's charisma. Dark complexioned people predominated in contrast to the stage where organizers, guests and personnel who were mainly of the upper- or the upper-middle-class sat. These people were light skinned, with blond or light coloured hair, well dressed, sporting audio-equipment, headsets, cell phones, and radios. Everything displayed enterprise efficiency and technology. The ethnic and socio-economic differences were very well-known to such an extent that one woman said to her companion: 'Wow! There is only pure white people', as she pointed to the stage with a tone somewhere between envy and submission.

The spectacle began at 6.30 pm. As the first invited politicians arrived, the entertainers shouted out slogans such as 'Spread the word, Fox has won', 'The one who will hound the PRI has arrived', and 'Down with the PRI'. Eventually, however, people became restless and called for Fox to appear. The entertainers and the team of security became nervous. At most they expected 100,000 spectators, not the 250,000 who were there. At this point the entertainers began to exercise authority. They demanded silence and calm. But the masses insisted: 'We want Fox! We want Fox!' The stage presenters were at a loss. They begged the multitude to change their spirits. 'Where are your smiles?' they asked in a worried tone. They were answered with: 'Get out, get out!' and 'Get them out!' Only when the speakers asked the sound engineer to raise the volume on the music were they able to silence the masses.

The Speech and the Masses

The way in which the speeches were presented and the response of the supporters during the electoral campaign were symptomatic of the political and civic culture that each political party and each political rally represented. These cultural characteristics are not an additional component of the political struggle. Instead, they represent the very essence of the political confrontation. Through them we come to know the essence of the collective behaviour and the meaning and application which is given to the public space. In Fox's meeting, people put up with the first presentations only because they signalled the beginning of the long awaited presence of the candidate. The crowd had to wait almost five hours for the rally to begin and they had already begun to feel anxious and tired. When Fox finally approached the podium it had grown dark. The lights and lasers on the giant screens and the advertising spectacle of the event were more spectaular in the dark. Fox was accompanied by two of his daughters. He was dressed in a navy blue suit altering his image from that of a sloppy candidate to a statesman. His speech took only 22 minutes. He directed himself to the people but tried mainly to persuade the political left. Indeed the most important implication in his speech was: 'Understand this, Cuauhtémoc! [his leftist opponent]. You are the only one missing'. Fox demanded much from his political adversaries. The political surveys predicted that at least a tie would emerge between him and Labastida of the PRI.

He offered a politically superfluous speech, though he was certain of political victory: 'I will work with all, even with the adversaries'. He elevated his voice to say: 'I will demolish the wall today, today, today!' The masses were transfixed by his speech and chanted 'Viva Mexico!' Just then fireworks illuminated the plaza. Even at 9.30 pm there was still great excitement. People refused to leave. They shouted, 'Put on your hat! Put on your hat!', hoping that he would put on the traditional Mexican hat. People liked the image of the Marlboro-style cowboy more than the look of a statesman. Fox's 'liabilities' – being a businessman turned politician, arrogant and conservative – became his assets amidst this crowd.[11] Many women were hysterical. Young people shouted and danced as if they were Fox's fan club. The national anthem was sung with great energy, emotion, and conviction. The candidate waved the national flag. Those present idealized him as strong and steadfast, as an idol. It was, without a doubt, a total media production. The PAN and its supporters used the public-political space of the Zócalo in a radically different way to the Zapatistas.

The March for Dignity

On 1 December 2000, barely five months after the PAN political rally, Vicente Fox would for the first time take control of the country. The mass media as well as citizens were joyous as, finally, democratic change was expected. That same day, however, in the Lacandona Forest, Subcomandante Marcos, the leader of the movement for indigenous rights of Mexico made a much less ostentatious declaration to the media. The EZLN was prepared to march to Mexico City represented by 23 commanders of highest rank of the Clandestine Committee of the Indigenous Revolution (CCRI). The aim was to promote the approval of the initiative for peace of the COCOPA.[12] This was in fulfilment of the San Andrés Accord after the dialogue concerning indigenous rights between the EZLN and the federal government in 1995, approved but later denied by the same government. The announcement sparked heated discussions around the country and the ensuing debate clarified the political positions of many groups, opening the wounds of class struggle in Mexico.[13]

The Conditions for Action and Public Debate

At the end of 2000, coverage of Fox's market-driven campaign for presidency began to dwindle, while Marcos's declaration gained more. Without doubt the aim of the march to Mexico City was to challenge Fox's popularity, and few failed to link it to the First Declaration of 1 January 1994 in the Chiapaneca city of San Cristóbal de las Casas where the EZLN announced their arrival before a stunned world. At that time, the neo-Zapatistas declared war on the federal government and the national army. They vowed to initiate a march to Mexico City. Now, 7 years had passed since the armed uprising. Would the Zapatistas, warriors against globalization and neo-liberalism, arrive physically and symbolically in the capital of the country? In political, financial, and security terms the March for Dignity required the assurance that civil-society organizations, public security officials, and the governors of the states through which they travelled would guarantee their safety. In the selection of the places where rallies were held, we see how space became an essential aspect of the Zapatista struggle: some places were defined by their civic and commercial significance; some were in the vicinity of traditional sites of domination of indigenous populations; some counted on the predominance of indigenous population; others had a tradition of social struggle; and some were linked to the life of the revolutionary Emiliano Zapata.

The Zapatista march, which lasted 36 days and covered 3000 kilometres, began as a symbol of national and indigenous dignity. It traversed 12 states where 33 political rallies were held; it gathered 3000 people in private vehicles and rented buses which followed the delegation along highways and travelled across the plains, valleys, and mountains of Mexico. This was a long and winding procession resembling an enormous serpent which, in the words of the Nobel Laureate José Saramago (1998), 'could not fit straight and therefore curved and coiled [and] was determined to arrive at all parts and to offer a spectacle to all cities and all places'.[14] At this stage, politicians began to polarize between supporters and denouncers, as governors and other institutional political representatives took sides. Ironically, the President astonished many by welcoming the march. Citizens' impression of Fox at the time was that of an honest and trusting president who was willing to receive the EZLN and willing to engage in dialogue. Few understood Fox's political strategies, especially his saving of the presidential image by delegating to political allies the denouncing of the Zapatistas.[15] In effect, Fox's allies talked about the situation in opposing terms. 'To march armed, with covered faces, seizing highways without doing anything to them is an affront to the people of Mexico', said PAN political representatives. According to them, the Zapatistas 'deserve[d] capital punishment'. Thus, this view contrasted with the position of President Vicente Fox who maintained an image of respect for plurality before the Mexican citizenry.

On the other hand, the representatives of the previous PRI government against whom the EZLN had declared war on behalf of indigenous dignity did not have a strong position and, at this time, appeared dismembered and leaderless. However, things were quite different with the social-democratic faction of the PRD. Some professionals and militants in this extremely diverse group supported the march – some with open declarations and resources and others through visible political activism. One important strategy that the rebels had assumed was establishing relations with organizations, intellectuals, and international politicians whose honesty and sincerity were unquestionable.[16] This allowed the EZLN to extend the public sphere they were part of. The marchers made several stops along the way to Mexico City so as to create symbolic ties to civil society. Mass media was used to influence public opinion and celebrated demonstrations with political and intellectual figures. This all constituted a type of political appropriation of the national territory as public space.

The results of the opinion surveys, however, were contradictory. Only 2 per cent of those interviewed knew that the Zapatistas wanted to meet members of the Congress. One-third of those polled thought that the Zapatistas went to the city with the objective of signing a peace accord. But this was not the case.[17] Certainly, citizens' sympathy towards the Zapatistas had increased as the march progressed, as two-thirds of the interviewees attested. The development of the march and its events were transmitted by radio, TV, and Internet to other countries of the world. It was evident that in the media struggle between Fox and Marcos the latter was gaining ground.

The Arrival in Mexico City's Public Space

The arrival on Thursday, 8 March 2001 was a symbolically loaded event. The Zapatistas were received by groups of very different types: student and youth associations, communities and farmer groups, workers, popular organizations and civil associations. Despite the great enthusiasm at the periphery of the metropolis, the masses wished to see Subcomandante Marcos enter the Zócalo. The 23 commanders along with Marcos sporting his characteristic smoking pipe and black knitted cap, were elevated onto the platform of an uncovered trailer, and travelled the streets and avenues. Figure 2.3 shows the symbolic appropriation of the main plaza. As indigenous Mexicans turned their backs to the National Palace, Marcos explained that 'It is no accident, it is because from the beginning, the government (must be) behind us'.

At the beginning, representatives of ethnic groups passed on containers of incense and copal in a ritual indigenous ceremony. Marcos's message was hardly belligerent; he did not claim to monopolize the truth. 'We do not come to tell you what to do nor to guide you to any side. We come to ask you, humbly, respectfully, to help us. That you do not allow the sun to rise without that flag that waves

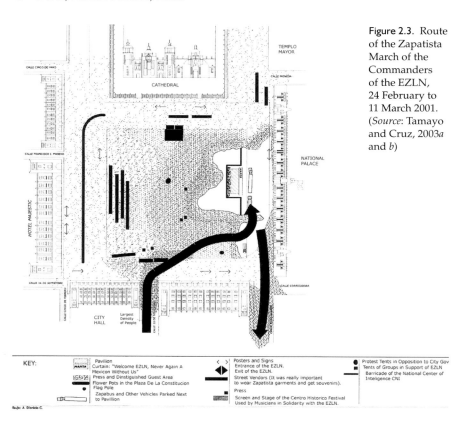

Figure 2.3. Route of the Zapatista March of the Commanders of the EZLN, 24 February to 11 March 2001. (*Source*: Tamayo and Cruz, 2003*a* and *b*)

imposingly in the centre of the plaza holding a worthy place for those of us whose colour is the colour of this land', he said as the 200,000 spectators cheered.

During the mobilizations in the city, spontaneous expressions of support on the part of housewives, children, the elderly, workers, and passers-by signalling victory were everywhere to be seen. One moment the presence of the commanders and Marcos quietened the crowd and the next their presence filled the crowd with excitement and deafening cheers: 'You are not alone, you are not alone!' 'We come to be heard. The congressional tribune is not theirs; it is property of all the Mexicans', replied the Zapatistas. The EZLN representatives made other appearances at the University City and before the legislature. Central to these great gatherings was the political struggle with the President of the republic, large-scale entrepreneurs, and the PAN. All the while, negative discussion grew concerning the EZLN speech at the congressional tribune. A dispute unravelled at this point. As the PRD called for openness and plurality, the PAN was steadfast in their opposition to the EZLN. As a consequence of intense lobbying, the commanders of the EZLN were allowed to enter the assembly of the House of Representatives. Four commanders and three leaders of the Indigenous National Congress spoke.

The immediate impact of the march was undeniably successful for the EZLN. President Fox capitalized politically, though apparently the event was also a defeat for the PAN and other conservative forces like a part of the Church and wealthy entrepreneurs. In effect, a few weeks later, the PAN and the PRI legislators would take revenge for this partial defeat and approve a law that opposed the previous agreements of San Andrés. The new law reflected the spirit of PAN ideals and was supported by a powerful group in the PRI. At this point, the President, who had previously boasted about his sympathies with the demands of the EZLN, kept silent at a key moment and in the process mended his differences with the PAN.

Conditions of the Public Space, a Theoretical Interpretation

In this section we reflect on the constitution of the public sphere in Mexico through the comparison of our two case studies. We begin by analysing the conditions, restrictions and opportunities for its existence. The conditions refer to the actors, the values, the mass media, and the physical space.

Actors and Citizen Practices

The liberal version of the public space assumes equality among individuals involved in dialogue, which is a conception far removed from reality. In the public sphere not all possess similar capacities for argumentation (as shown in figure 2.1). Liberal notions such as this are blind to the reality of social classes, and cultural and ethnic heterogeneity. However, we can speak of the existence of individuals (citizens) who struggle to secure their rights. It is the inequality of endowments and not the equality of conditions that generates social struggle. This understanding of communication is not a property of the public sphere as defined by Habermas. Instead, it corresponds to the empirical understanding of public space. Building on Honneth and others, if we are to understand the discursive struggles for cultural and political hegemony, any analysis of public space requires the factoring in of the role of social groups who are dominated as much as the contours of social conflict.

As such, a component of public space is the history and experience of citizenship.[18] And citizenship, like the public space, is an arena of conflict (Alejandro, 1993). Following Durkheim's theoretical tradition, citizenship is a legitimization of the social contract between diverse individuals. It is for this reason that it is endlessly reconstituted, while the state seeks to influence what are conceived to be legitimate definitions. It attempts to set the parameters of citizenship and public space, and to impose a meaning and a direction on their dynamics. However, the public space created in both case studies was characterized by a diversity of actors and interactions: citizen groups, political factions, leaders, organizers, political parties, etc. Diverse networks and alliances with unequal levels of political weight were formed. They interacted based on their objectives, motivations, and interpretations of the conflict that they themselves were staging.

We began our empirical analysis with an understanding that the generated space was a field of confrontation between the actors in civil society and of associated individuals and social movement organizations (see figure 2.4). The public space, like the electoral campaign and the indigenous people's march, was redefined with actions and symbolic forms. We refer not only to the will and the expressed attitude of actors to establish either a dialogue or a dispute, but to the political necessity for some groups to engage in such activities. This was expressed in the first case through the political parties, and in the other case through a social movement.

As has been demonstrated above, neither citizenship nor public space are neutral (Tamayo, 1999, 2002). Public space is not a homogenous expression of the general will either. It is instead a space for struggle, a site of past memories, present experiences, and perspectives on the future.

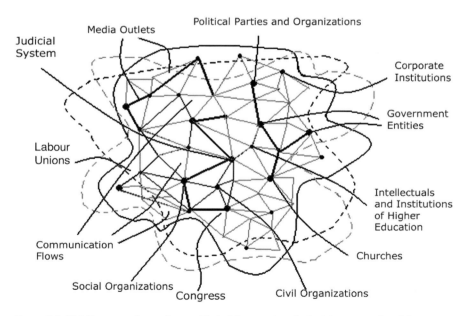

Figure 2.4. Public space. Space is constituted by a network that is appropriated by actors with different positions and values. Actors are related through communicational flows with different intensities.

Values

Values here refer to the virtues and the meanings of the public sphere, e.g., freedom, reason, individual autonomy, justice, norms of debate in order to reach consensus, the clarity of speech, veracity, sincerity, etc.[19] However, these values depend on the definition that political actors give them. At this point we emphasize three necessary values: freedom, justice, and autonomy. For Arendt, collective action is not sufficient to delimit the space of the public sphere. It is fundamentally important to imbue space with value. That value is the value of freedom that grants independence, autonomy, and emancipation to human beings. Rawls would add the value of justice, understood as fairness. For Habermas, autonomy is an extension of the individuals' capacity to communicate in a process of rational deliberation. Accordingly, individuals participate to the extent that they take part in a reflexive and public debate. Consequently, the simplest valued conditions for the establishment of the public space would be: 1. that private individuals have the freedom to make public use of their reason, i.e., to make public their private reasoning; 2. that private individuals are autonomous, with their own capacity to participate; 3. that state power is put under public judgment; and 4. that this judgment be the fruit of consensus.

But who are the carriers of these values? Are they all those who participate? How can we distinguish them? From the speech and the rules of the political game in both case studies we found that delegates were representatives, leaders, and spokesmen for social groups who were able to express themselves and to make use of the power of discourse to organize and to participate. However, those who were authorized to question or to present/display ideas did not speak for themselves as much as they were representatives of established ideals and values.[20] Nevertheless, freedom, justice, and autonomy were not values implicitly understood by the actors. Instead, each actor interpreted values according to his or her own position in the public space. In the case of the Zapatistas, the freedom of this group to arrive in Mexico City was opposed to the ideology of the PAN, whose members had declared that snipers would be ready to receive them. In this way, the public sphere once again became a battlefield.

It is on the basis of these communicative actions and interactions that collective identities were constructed and reinforced. Different groups and adversaries identified themselves and confronted others in the same space but in asymmetric places, as they created hegemonic speeches. Their discursive identities were constructed through the interactions and the content of speeches. The essential values of individuals were defined by the symbolic forms of discourse and action, political pressure, and political activity.

The normative principles that are derived from our analysis have been characterized first by the identification of deliberation as a process (in the electoral campaign and the Zapatista march, the degree of freedom to discuss was not determined by the acceptance and toleration of the adversary but rather by the force of discursive imposition). Secondly, they have been characterized by the degrees of inequality in the participation of the actors in struggle (in the case of the electoral struggle, power differentials depended on the resources, on the mobilization, and on the historical experience of its exercise; and in the case of the indigenous Mexicans, in the composition of the actors: poverty and political marginalization by the omnipresent institutional powers of the legislative and executive government). Thirdly, they are characterized by the possibility of reaching a minimal level of agreement among citizens who can establish a general will (this however was mediated, at least in the case of the Zapatistas, by the existence of diverse actors who used violence and/or the threat of political pressure). Significantly, in each one of these normative principles, the important factor is not the liberal ideal of free and just deliberation, or the pre-existence of a general will. Instead, the fundamental aspects are the forms, degrees, and pressures in which these processes of deliberation and struggle for power occur.

Media

In reframing the idea of public space, Habermas (1993) recognizes the transformation that the public sphere has had in history. Within this perspective, there are two fundamental transformations: change with regard to the state and civil society, and change in communicative structures. The transmutation of mass media is based on impressive developments in science and technology. Historically, it has been the democratic revolutions in France and the United States, mainly, but also the experience of wars for independence in Latin America and the liberal democratic victory that have facilitated the institutionalization and expansion of the public sphere in society (Annino, 2003; Hamnett, 2003, Palti, 2003). The media have been transformed and have contributed to a discourse that has impacted the formation of certain categories of citizenship in each country. Indeed, Gingold (2000) emphasizes the fact that public opinion does not occur through expensive face to face communication, but through multiple communicative flows. The media, which are supposed to facilitate communication, have come to define communication and in the process have manipulated individual and collective reasoning. However, it is important to recognize that in spite of the deep technological changes in communication technology, there exist communication media that have remained seemingly unaltered, as well as others that have been transformed in history. Newspapers and literary publications, for instance, coexist with newer, high-technology media.

Also the concept of publicity has changed with the development of mass media. Today, publicity is related to commercialized media and market-driven information that are oriented to induce consumption and not political discussion of ideas. The development of the media, the increased size of society, and commodification have altered the direction of publicity away from politics. This denigration of media production into spectacle fractured the link between the public sphere and the individual, as Habermas has asserted. Specifically, technology has changed the ways in which information is transmitted. Speakers are increasingly able to interfere with the interpretation and manipulation of news, and media have facilitated the selective emphasis of certain types of information, furthering distinct ideological interests.

In spite of this, we cannot speak of a passive unreflective citizen, since public discussion proceeds at the formal and the informal levels. Communicative action is furthered by informal political discussion that at times is sustained by gossip and rumours. All this has shaped the social ideologies and imaginaries. Formal and informal media have included partisan newspapers, diverse publications, parliamentary debates, assemblies in political clubs, entertainment clubs, business organizations, and meetings in social organizations. The space of the Zócalo has facilitated the production and distribution of messages for moulding political discourse.

The Physical Space

In the two cases studied, physical space was a condition of the public sphere. Space was thus an indispensable tool for exercising citizen rights (Claval, 2001). Its physicality became the context within which groups of different ethno-racial and cultural categories and individuals from different economic classes encountered one another (Ghorra-Gobin, 2001b). As Chelkoff and Thibaud (1992) have mentioned, social interactions take place through environments configured by their defining objects and surrounding landmarks. The relations of reciprocity among actors and between them and the structured environment were the public scene.

Some geographers and city planners have criticized the ambiguity of the term 'public', and blame social scientists for manipulating the term 'space' when discussing the public sphere.[21] Alternatively, scholars of the public sphere have criticized certain positions in urban studies that reduce the concept of the public sphere or public space to very practical notions of open urban space for every day use. The vision of public space as simply that which is available for the masses to occupy and circulate freely within is incomplete. This incomplete understanding does not consider some fundamental conditions of space either as public or private property.[22]

The liberal concept of urban public space according to Blanc (2001) has become too ambiguous – it is anything and everything: street, television, coffee shop, Internet, etc. However, the physicality of space disappears with such an idea of public space, but, we assert, physical space has not disappeared, though it has been transformed. Its transformation, however, is rooted in contemporary technology and modern forms of the relation between state and society, which transform the urban cultural experience. Urban space suffers from the impact of the transformations and crises of the city. The streets, plazas, and meeting places are also affected as a result of cyclical economic recessions, the impact of globalization on local space, and the deepening of poverty and social segregation (Tamayo and Wildner, 2002). But as the structural crises of capitalism do not necessarily imply its collapse, streets and plazas also persist but transform.[23]

We have seen the public sphere formed by different groups in constant tension and crystallized in concrete places. In the symbolic appropriation of the city by indigenous Mexicans, an enormous network of places was constructed that included residential districts, factories, work centres, plazas and patios in the inner parts of the buildings, farms, community houses, bars, street corners, etc., where rumour and gossip were discussed (as illustrated in figure 2.5). Of course, it is important to recognize that none of these locations is naturally a space for

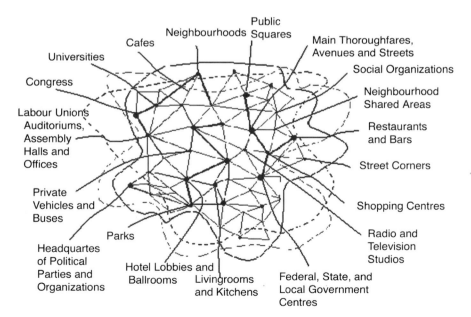

Figure 2.5. Public Space. Space constituted by formal and informal places, and some are occasionally turned public by the actors. They are connected by access routes.

public debate. Instead, social actors have fashioned those places for discursive exchange. Tomas (2001*a*) criticizes certain tendencies in social science that overlook concrete space where public debate is crystallized. The political analysis of the public sphere in the electoral campaign finale and the indigenous Mexican march are strengthened when the importance of physical space is recognized. The spatialization of the public draws particular attention to the forms of political (and physical) appropriation of urban space.

Following Tomas, the electoral campaign finale and the indigenous march demonstrate that public space is not waning.[24] It is through these examples that we see how public space manifestations were beyond the idealization of the Greek agora or the Roman forum. It is not enough to only describe the interactions of the individuals in physical space, since the flows of the political debate can quickly go beyond their physical parameters. The public space in Mexico City has experienced similar transformations to those of the public sphere: it is a place for face-to-face encounters and re-encounters which has been modified but not disappeared. In those places residents and citizens discuss, rejoice, lament, walk in public, and hold an affinity for their historical patrimony (Tomas, 2001*a*). Public space can be a site for consensus as much as a site of conflict or of political confrontation.

Final Considerations

Our intent in this chapter has been to link physical, social, and symbolic space with a notion of political space through the cases of the 2000 electoral campaign and the Zapatista March. We stated that physical space is not public solely because it is a place where people meet freely. It is also public because of the intellectual and political consequences of those encounters. It is public because of the ways in which space is appropriated. In an illustration of the bonds that exist among public space, social practice, and political conflict we made use of a critical vision of the concept of the public sphere. We described two extraordinary situations in the same place and attempted to understand how such events reinforced the conditions of the public sphere in Mexico.

We emphasize the importance of recovering physical space as a fundamental condition of the public sphere. We have argued against a culturalist vision of everyday use of public space as it displaces the perspective of conflict, of inequality, and politics. We have also criticized the political vision that displaces those cultural practices that take place through the appropriation of physical space. We have argued that to separate one from the other hinders a holistic view of reality. Consequently, it is the exercise of citizenship that gives public space its specific and historical character. In the same manner, we have indicated that there are no existing values which are naturally tied to the public sphere. It is the media at the formal and informal levels, as well as the social and political actors, who define and redefine those values.

The extraordinary events of the electoral concentrations and the presence of the Zapatistas in a place like the Zócalo of Mexico City deepened the symbolic and political character of this urban space. Yet, at the same time, the historical symbolism etched into that place offered such events a more extraordinary character. Indeed, this can be traced in the Zapatista struggle to create an alternative discourse. The meaning of those events is irreducibly tied to the images and meanings of space in the memory of the citizenry. The political appropriation of the public space becomes an essential condition for the construction of the public sphere.

Notes

1. See Bolos (2003), Ramírez Kuri (2003), Olvera (2003), Avritzer (2002), Low (2000), Panfichi (2002), Dagnino (2002), and Alvarez (1997).
2. The author nevertheless clarifies that collective action is not sufficient to delimit the space of the public sphere. To become a political project among equals, it is necessary that individual freedom is protected.
3. Habermas's writings, although not sufficiently representative, are mandatory for the study of public opinion (Gingold, 2000; Tomas, 2001a; Sennett, 1974). The colonization of the lifeworld is interpreted as the dominion and authority of power and money (the system) on the world of the living (lifeworld). To us, public space serves as a link between the system and the lifeworld.
4. Conversely, Coutterau and Quéré (2003) question the notion of different public spheres. Conditions such as publicity, public place, the forms of public management, and what McAdam, Tarrow and Tilly (2003) call contentious politics reconfigure different 'spheres' in the same 'space'.
5. For Smelser, social movements can have normative as well as value-based prerogatives. On the one hand, normative goals attempt to reform regulations and institutions; on the other, value-based goals are oriented more towards questions of the legitimacy of the system. For Smelser, ideologically motivated movements that aim at political revolution prevail. In Mexico, the electoral struggles which have taken place mainly within the framework of the democratic transition are motivated by normative reasons. The indigenous fight of the EZLN is, on the other hand, as much value-based as it is normative-based, since they both question the legitimacy of the Mexican system and request constitutional reform.
6. It is necessary, nevertheless, to consider the dimension of dialogue and action as indicated in Arendt's work, just as much as one would consider the dimension of understanding and interpretation argued by Habermas (Honneth, 2000).
7. Polls gave Fox (PAN) 41.7 per cent, Francisco Labastida (PRI) 37.7 per cent, and Cuauhtémoc Cárdenas (PRD) 17.6 per cent (Tamayo, 2002).
8. See Tamayo (2002) for an ethnographic study of the interaction between social and physical space, a hermeneutic analysis of agents' interpretations, and news reports and documentation of the political and urban context.
9. These landmarks included the Angel of Independence, the traffic circle at Chapultepec, Avenida Reforma and its postmodern buildings, the Sanborns Cafés and the rotundas of the Diana Cazadora sculpture, and the monument of Cuauhtémoc, the last Aztec emperor.
10. The PAN had been a party of the middle class, with professionals and small

businessme　　　base. After the conflicts between big business and the state in 1976, the PAN received the political support of other prominent businessmen who leaned towards the right (see Tamayo, 1999).

11. Other Fox epithets were 'Coca-Cola businessman', 'authoritarian', 'pragmatic', 'evasive', 'devout Catholic', 'directly tied to the international Christian Democratic party of the rightist president of Spain, Jose Maria Aznar', 'cowboy', 'businessman turned political strongman', 'graduate of the Ibero-American Jesuit University', etc. (see the campaign dossier of Vicente Fox from January 1 to July 2, 2000).

12. The Commission for Peace and Cooperation in Chiapas (COCOPA) was created by the Congress of the Union in an attempt to contribute to peace in Chiapas. It had 12 deputies, 6 senators and 1 representative of the Local Congress of Chiapas. The positions of president and spokesman of COCOPA were rotating.

13. In this section we synthesize daily news sources, chronology of events during the arrival of the 24 representatives of the EZLN to the Zócalo in Mexico City, articles concerning the event in different national newspapers and periodicals, and analyses of the political, social, and spatial impacts (see also Tamayo and Cruz 2000*a* and *b*).

14. José Saramago provides this vivid description in his book, *Memorial del Convento* (1998).

15. President Fox had no intention of reaching a peace agreement and apparently neither did Marcos. However, Fox called for respect to the caravan of the commanders of the Zapatistas and qualified it as a 'march for peace'. On Friday, 23 February 2001, one day before the beginning of the march, Fox spoke on national TV and welcomed the march to Mexico City.

16. The discussion included representatives of the European Parliament, the Italian Monos Blancos and the Comité de Presos y Exiliados Políticos, as well as poets, activists, social writers and scientists. It was important to generate such alliances in order to resist the support that European and American institutions gave President Fox.

17. The statistical data on public opinion in this section are based on national surveys that were conducted on 17 February, 3, 4 and 11 March 2001. Published in the *Reforma* Newspaper 23 February and 7, 12 and 29 March.

18. Citizenship is a multidimensional and dialectic relationship between the state and civil society. This multidimensional relationship is in essence conflictual, and these conflicts render historical character to public space (Tamayo, 1999).

19. Habermas denotes the set of values that justify the existence of the public space as the Ideal Speech Situation. In fact he refers to the conditions for a true dialogue to take place.

20. It is necessary to note that the public sphere of the eighteenth and nineteenth centuries is no longer valid.

21. In several works, particularly those in France, a mistake has been made in tying the concept of public space to Habermas, given the publication of his book originally titled *Strukturwandel German to der öffentlichkeit* (Structural Transformation of the Public Sphere) in 1962. The problem is that it was erroneously translated as *L'espace public* (Public Space) in French, when in fact the author wanted to talk exclusively about the formation of the bourgeois public sphere in terms of communicative action.

22. For this reason it is not possible to define private property, like commercial centres for example, as public. The collective use that is made possible in commercial centres is restricted as access is controlled and proprietors reserve the right of admission. For further discussion of these subjects see Tomas (1998), Gottdiener (1995), Wildner (2005), Portal (2001), Caprón (2001), De la Pradelle (2001), Urteaga and Cornejo (2001), Cornejo (2001), Ramírez Kuri (1993); for the appropriation of public space, see Wildner (1998; 2005) and Monnet (1995). The workshops of Urban Ethnography organized by the Metropolitan Autonomous University also studied commercial centres as sites of globalization. For additional information see Tamayo (1999) and Ortiz and Tamayo (2001).

23. For Richard Sennet (1974) 'the death of the public man' is observed in all the levels of the social life, in residence, in work, and in the cultural realm.

24. Although Jane Jacobs (1961) and Richard Sennett (1974) have warned against the danger of the disappearance of public space and the fall of the public man, another important risk is the permanent imposition by powerful minorities of their wishes on majorities.

References

Alejandro, Roberto (1993) *Hermeneutics, Citizenship, and the Public Sphere.* New York: State University of New York Press.
Álvarez, Lucía (1997) El proceso participativo y la apropiación del espacio público en el Distrito Federal, in Álvarez, L. *Participación y democracia en la Ciudad de México.* Mexico: La Jornada Ediciones.
Annino, Antonio (2003) Definiendo al primer liberalismo Mexicano. *Metapolítica*, **7**(31).
Arendt, Hanna (1961) *Condition de l'homme moderne.* Paris: Calmann-Lévy.
Avritzer, L. (2002) *Democracy and the Public Space in Latin America.* Princeton, NJ: Princeton University Press.
Blanc, Jean-Nöel (2001) Voir l'espace dans l'espace public. *L'espace public. Géocarrefour. Revue de Géographie de Lyon*, **76**(1).
Bolos, S. (ed.) (2003) *Participación y Espacio Público.* Mexico: Universidad de la ciudad de México.
Caprón, G. (2001) De la 'marchandisation' del espace public: le cas du centre commercial à Buenos Aires, in Ghorra-Gobin, C. (ed.) *Réinventer le sens de la ville: les espaces publics à l'heure globale.* Paris: L'Harmattan.
Chelkoff, Grégoire and Thibaud, Jean-Paul (1992) L'espace public, modes sensibles. *Les Annales de la Recherche: Les Espaces Publics en Villes.* Nos. 57–58.
Claval, Paul (2001) Clisthène, Habermas, Rawls et la privatisation de la ville, in Ghorra-Gobin, C. (ed.) *Réinventer le sens de la ville: Les espaces publics à l'heure globale.* Paris: L'Harmattan.
Collin, Denis (2001) *Morale et justice sociale.* Paris: Editions du Seuil.
Cornejo, I. (2001) Plaza Universidad: ¿Estar en un centro comercial es una manera de hacer ciudad? in Aguilar, M.A., Sevilla, A. and Vergara, A. (eds.) *La ciudad desde sus lugares: trece ventanas etnográficas para una metrópoli.* Mexico: Miguel Ángel Porrúa, CONACULTA, UAM Iztapalapa.
Coutterau, A and Quéré, L. (2003) Postface, in Barril, C. *et al.* (eds.) *Le public en action. Usages et limites de la notion d'espace public en sciences sociales.* Paris: L'Harmattan.
Covarrubias, Ana Cristina (2000) Encuestas y elecciones: primeras evaluaciones. *Este País*, No. 113, p. 42.
Dagnino, Evelina (ed.) (2002) *Sociedad civil, esfera pública y democratización en América Latina: Brasil.* Mexico: FCE.
De la Pradelle, M. (2001) Espaces Publics, espace marchands: du marché forain au centre commercial, in Ghorra-Gobin, C. (ed.) *Réinventer le sens de la ville: Les espaces publics à l'heure globale.* Paris: L'Harmattan.
Foucault, M. (1994) Espacios diferentes, in *Toponimias (8) ideas del espacio.* Madrid: Fundación la Caixa.
Ghorra-Gobin, C. (2001*a*) Les espaces publics, capital social. *L'espace public. Géocarrefour. Revue de Géographie de Lyon*, **76**(1).
Ghorra-Gobin, C. (2001*b*) Réinvestir la dimension symbolique des espaces publics, in Ghorra-Gobin, C. (ed.) *Réinventer le sens de la ville: les espaces publics à l'heure globale.* Paris: L'Harmattan.
Gingold, Laura (2000) Opinión Publica, in Baca, L. *et al.* (eds.) *Léxico de la Política.* Mexico: FCE.
Gottdiener, M. (1995) *Postmodern Semiotics, Material Culture and the Forms of Postmodern Life.* Oxford: Blackwell.
Habermas, J. (1993) *L'espace public* (translated by Marc B. de Launay). Paris: Payot.
Hamnett, B. (2003) El liberalismo mexicano del siglo XIX: Origen y desarrollo. *Metapolítica*, **7**(31).

Honneth, A. (1996) La dynamique sociale du mépris. D'où parle une théorie critique de la société? in Bouchindhomme, Ch. and Rochlitz, R. (eds.) *Habermas, la raison, la critique*. Paris: Cerf.

Honneth, A. (2000) *La lutte pour la reconnaissance* (translated by Pierre Rusch). Paris: Cerf.

Jacobs, Jane (1961) *The Death and Life of Great American Cities*. New York: Random House.

Low, Setha M. (2000) *On the Plaza. The Politics of Public Space and Culture*. Austin, TX: University of Texas Press.

McAdam, D., Tarrow, S. and Tilly, Ch. (2003) *Dynamics of Contention*. Cambridge: Cambridge University Press.

Monnet, J. (1995. *Usos e imágenes del Centro Histórico de la ciudad de México*. Mexico: DDF.

Olvera, Alberto J. (ed.) (2003) *Sociedad civil, esfera pública y democratización en América Latina*. Mexico: Universidad Veracruzana y el Fondo de Cultura Económico.

Ortiz, J. and Tamayo, S. (2001) Metodologías cualitativas en la enseñanza del diseño: arquitectura y espacios urbanos, in *Anuario de Estudios de Arquitectura, historia, crítica, conservación, 2001*. Universidad Autónoma Metropolitana, Azcapotzalco, pp. 9–25.

Palti, Elías J. (2003) El pensamiento liberal en el México del siglo XIX: Trascendencia e inmanencia. *Metapolítica*, **7**(31).

Panfichi, Aldo (ed.) (2002) *Sociedad civil, esfera pública y democratización en América Latina: Andes y Cono Sur*. Mexico: FCE.

Portal, M. (2001) Del centro histórico de Tlalpan, al centro comercial Cuicuilco: la construcción de la multicentralidad urbana, in Aguilar, M.A., Sevilla, A. and Vergara, A. (eds.) *La ciudad desde sus lugares: trece ventanas etnográficas para una metrópoli*. Mexico: Miguel Ángel Porrúa, CONACULTA, UAM Iztapalapa.

Ramírez Kuri, P. (ed.) (2003) *Espacio Público y reconstrucción de ciudadanía*. Mexico: Miguel Angel Porrua y FLACSO.

Ramírez Kuri, P. (1993) Transformaciones espaciales y modernización urbana. La ciudad de México y los macroproyectos comerciales: Centro Comercial Coyoacán. Tesis de Maestría en Estudios Regionales. Mexico: Instituto Dr. José Luis Ma. Mora.

Saramago, José (1998) *Memorial del Convento*. Mexico: Alfaguara.

Sennett, Richard (1974) *The Fall of Public Man*. New York, NY: Knopf.

Smelser, N. (1995) *Teoría del comportamiento colectivo*. Mexico: Fondo de Cultura Económica. (first published in English in 1963).

Tamayo, S. (1999) *Los veinte octubres mexicanos. Ciudadanías e identidades colectivas*. Mexico: Universidad Autónoma Metropolitana, Azcapotzalco.

Tamayo, S. (2002) *Espacios Ciudadanos, la cultura política en la ciudad de México*. Mexico: Uníos y Frente del Pueblo, colección Sábado Distrito Federal.

Tamayo, S. and Cruz-Guzmán, X. (2003*a*) La marche de la dignité indigene. *Le Mouvement Social*, no. 202.

Tamayo, S. and Cruz-Guzmán, X. (2003*b*) Urban imaginaries and perceptions of the EZLN in Mexico City. *Monopolis: Globalisierung & Stadtforschung* (Globalization and Urban Studies), Sinn-haft 14/15. Vienna: Löcker Verlag.

Tamayo S. and Wildner, K. (2002) Lugares de la globalización. *Revista Memoria*, No. 156.

Tomas, F. (1998) ¿Y después del funcionalismo qué? in Tamayo, S. (ed.) *Espacios Urbanos, Actores Sociales y Ciudadanías*. Mexico: Universidad Autónoma Metropolitana, Azcapotzalco.

Tomas, F. (2001*a*) Du centre civique à l'espace public. *L'espace public. Géocarrefour. Revue de Géographie de Lyon*. **76**(1).

Tomas, F. (2001*b*) L'espace public, un concept moribond ou en expansion. *L'espace public. Géocarrefour. Revue de Géographie de Lyon*, **76**(1).

Urteaga, M. and Cornejo, I. (2001) Los espacios comerciales: ámbitos para el contacto juvenil urbano, in Aguilar, M.A., Sevilla, A. and Vergara, A. (eds.) *La ciudad desde sus lugares: trece ventanas etnográficas para una metrópoli*. Mexico: Miguel Ángel Porrúa, CONACULTA, UAM Iztapalapa.

Voirol, O. (2003) L'espace public et les luttes pour la reconnaissance. De Habermas à Honneth, in Barril, C. *et al.* (eds.) *Le public en action. Usages et limites de la notion d'espace public en sciences sociales*. Paris: L'Harmattan.

Wildner, K. (1998) El zócalo de la ciudad de México. Un acercamiento metodológico a la etnografía de una plaza, in *Anuario de Espacios Urbanos 1998*. México: Universidad Autónoma Metropolitana, Azcapotzalco.

Wildner, K. (2005) *La plaza mayor: ¿Centro de la Metrópoli? Etnografía del Zócalo de la ciudad de México*. México: Universidad Autónoma Metropolitana.

Chapter 3

Reinventing the Void:
São Paulo's Museum of Art and
Public Life along Avenida Paulista

Zeuler R. Lima and Vera M. Pallamin

An Extraordinary Void

In the middle of the dense, busy, and sharp-edged urban landscape of Avenida Paulista in São Paulo, there is a special open space that has captivated the collective imagination, emotions and actions of several generations of the city's population. This place is Trianon Terrace. It began as a park and now lies as a plaza beneath the wide concrete structure of the Museum of Art of São Paulo (MASP), designed by architect Lina Bo Bardi between 1957 and 1968. Despite remaining open for more than a century, or perhaps because of this, the terrace has been a unique stage for the urban and social development that transformed a small village in the late nineteenth century into one of the largest contemporary metropolises in the world.

According to Lina Bo Bardi, the Italian-Brazilian architect whose provocative ideas stirred up architectural debate in Brazil in the second half of the twentieth century, the shape of MASP did not result from some kind of 'architectural extravagance' (Bardi, 1994). Instead, she insisted that it resulted from the limitations imposed by the site. Since the creation of Avenida Paulista as a residential subdivision in 1891, the area currently occupied by the museum was reserved for a park, framing a picturesque view of the north side of the city. A law dating from the subdivision prohibited the building of any structure that would obstruct this panorama. In the 1910s, the city built a large terraced and ornate structure of pergolas and a hall, which was demolished in the 1950s. Nearly 20 years later, the museum reframed the large void of the plaza and the urban vista with a simple but bold architectural gesture.

Lina Bo Bardi's response to the legal restrictions created an unusual building configuration: a large glass prism that seems to hover above the vast empty space of the terrace. Rising from two reflecting pools, two red concrete piers spanning almost 240 feet hold the transparent volume above the terrace. Lina Bo Bardi used the pre-stressed concrete structure to create a simple and unpolished urban landmark. The museum opened in 1969, giving the city an unexpected open public space and reinforcing its reputation as a key cultural venue in São Paulo. The official designation of the plaza, Trianon Terrace, refers to its origin as a park, but it has not survived as a popular name. It is commonly known as *vão do MASP* (span of MASP), a term that sounds both simple and defying, both ordinary and extraordinary. Lina Bo Bardi referred to this space as the space of freedom in reaction to the increasing censorship imposed by the military regime at that time. Historically, the uses of the plaza have confirmed her aspirations.

The space beneath the museum was there from the start as an idea and as an image. It existed as a potential space long before its designers and its users could give it the different meanings it has acquired over time. In this process, it became, together with the museum, one of the most important open spaces of collective reference in the social and cultural life of São Paulo. Sitting across from Siqueira Campos Park (named Trianon Park in the past), the span of MASP realizes the aspirations of democratic civic life that permeated architectural and urban discourses after World War II. The purpose of the plaza designed by Lina Bo Bardi and its everyday uses resonate with the Eighth CIAM (Congrès International d'Architecture Moderne), which proposed the search for the 'heart of the city'. As the city grew vertically and horizontally, it incorporated the terrace into a dense fabric with few open public spaces. As a result, the plaza became more a focal point in the city than an outlook on to the surrounding landscape. As the vista reversed inwards, the terrace became a key venue for staging social, cultural, political, and economic activities in São Paulo.

The span of MASP has become a place of high visibility and strong political reference in the city. The genealogy of Trianon Terrace as a collective urban space is intertwined with the complex history of Avenida Paulista and also with the development of the city at large. It is at the same time the window on and the image of collective life in São Paulo. It is a space that has been continuously reconstituted by spontaneous and organized forms of public and private use, revealing different conceptions of collective urban life and social and political imagination.

Everyday life has historically accompanied extraordinary events along the Avenida Paulista, moving from the emergence of a local capitalist elite in the turn of the twentieth century, to the establishment of an urban middle class in the middle of the century, and to the multiple claims of citizenship and political and social participation in the last decades. The particular changes in the meaning of Trianon Terrace show how public administrations, civil society and the citizenry of São Paulo have imagined, claimed, occupied, and transformed referential urban open spaces in the city. This is the transformation we will analyse in the following pages.

Ordinary Urban Life and the Spaces of Dissent, Otherness, and Citizenship

The social, cultural and political appropriation of Trianon Terrace offers significant examples for thinking about how urban spaces are intimately related to the exercise of democracy, the claim for the political recognition of otherness, and the redefinition of the role of citizenship in contemporary Western societies. Negotiations for social belonging and public participation as well as struggles for social and moral recognition have increasingly chosen the span of MASP as a stage in the metropolis. They operate according to what political philosopher Jacques Rancière (1996) described as the practice of dissent.

The term dissent, or disagreement,[1] was proposed by the philosopher to enhance the role of difference and to work out social antagonism and cultural variations in the public sphere. This notion helps us understand and to mediate the definition of urban public and private spaces, as well as the interests, individuals and groups included and excluded from them. Contrary to the notion of consensus, dissent relates existing conflicts in the constitution of what is visible, pronounceable, or feasible in such spaces. The confrontation between different voices does not necessarily need to be seen as a negative relationship of opposition. According to Rancière (1996, p. 374), it does not have to be 'a war of all against all'; it promotes ordered situations of conflict, and encourages discussion with the inclusion of groups that have been either silenced or excluded from the exercise of citizenship and from the participation in the constituency of public places and the public sphere.

Dissent is a democratic means that calls into question the assumed consensus of modern rationality. Instead of proposing universal meanings, it fosters continuously renewed and inclusive social debates. Dissent rearticulates modern categories such as identity and national and cultural citizenship in a contemporary world that calls for the recognition of difference and promotes the expansion of cultural repertoires reshaped by the expansion of global capitalism. The logic of dissent translates the transformation of social subjects into political interlocutors, and allows for the creation of political space and discourse. The openness to include those who have not participated in the debate creates new rules and new relationships. In the spaces of the city, new and old political subjects have to constantly reinvent themselves at the same time as they reinvent the city and the norms of the debate.

According to Rancière, this logic of participation is the logic of otherness. It embraces different sensitive worlds and constitutes a relationship shaped by opposition in the affirmation of individual and social identities. This relationship produces spaces of difference as controversial and political constructs, enlarging the traditional definition of citizenship from the sphere of the nation-state towards the sphere of the political, social, and lived space of cities. As the void under the Museum of Art of São Paulo transforms over time, the citizenry of São Paulo advance the contested history of inclusion and exclusion in the formal and informal practices of urbanization and in the public life of the city.

Creating a Void: Trianon Terrace in the Urban Development of São Paulo

The origin of the terrace beneath the Museum of Art of São Paulo is related to the creation of Avenida Paulista in 1891. This is an origin of exclusiveness. The neighbourhood along the avenue was the most fashionable and desirable residential development in the turn of the twentieth century, defining its symbolic presence in the social life of the city and strengthening the spatial division between affluent and working classes in the urban development of São Paulo. Uruguayan agronomist and developer Joaquim Eugênio de Lima conceived the avenue and a few adjacent streets as a luxury subdivision, catering to the emerging elites that connected Brazil to the international coffee trade. He acquired a large extension of lots from small rural properties beyond the southern outskirts of the city skirts, next to Caaguaçu Forest and along the ridge between the two major river valleys that became the geographic basis of the contemporary urban settlement. The original configuration of almost 3 kilometres of Avenida Paulista was based on design principles that privileged views of the surrounding topography, including a wide perspective of the historic centre of the city framed by Trianon Park. This vista belonged to the lot that later became Trianon Terrace (Lima de Toledo, 1987) as mentioned above. The subdivision reproduced the aesthetic and infrastructural patterns of European and North-American residential boulevards, including large lots with big mansions, manicured greenery and trees along wide streets and sidewalks, a small park, sewer system, and public transport.

São Paulo went through its first burst of economic and social development in the period between the early 1870s and late 1920s. The city occupied a strategic position between the areas of agricultural production in the hinterland of the state and the port city of Santos 50 kilometres south-east on the coast. This location made trade easy and accumulation of wealth and rapid urban growth possible in the final decades of the nineteenth century. The population of the city went from 19,347 in 1872, to 64,934 in 1890, and to 279,000 in 1905 (Sevcenko, 2000, p. 78). New urban residents from the large influx of immigrants constituted a massive working class, which contributed to the development of unions and political insurgence,[2] while some eventually became merchants, professionals, investors, and industrialists. By the 1920s, the population of São Paulo reached 579,000, and the city became the national leader in commercial and financial activities as well as in industrial production. Urbanization expanded at unprecedented speed, but investment distribution was uneven and infrastructure and public services were deficient in several new developments.

In the meantime, the city government together with private investors hired European architects and urbanists such as Joseph-Antoine Bouvard, Alfred Agache, Paul Villon, Barry Parker and Robert Unwin to embellish the city and also to design subdivisions in expanding affluent areas. Real estate interests were often an obstacle to urban master plans, concentrating public investments in selected neighbourhoods. In the decades to come, this form of urban growth contributed to the general spatial separation between wealthy and working classes in the south-west and north-east regions of the city, respectively. São Paulo is today a

Figure 3.1. Avenida Paulista in the beginning of the twentieth century. (*Photo*: Guilherme Gaensly. *Source*: Archive of Departamento de Patrimonio Histórico, São Paulo City Hall / DPH–PMSP)

metropolis punctuated by a few areas with great vitality and a high concentration of infrastructure, such as Avenida Paulista, and large areas of poverty and a severe lack of urban resources. This spatial division represents a long process of social inclusion and exclusion from the rights of citizenship, which has been more and more contested in the last few decades.

The original urban amenities created along Avenida Paulista marked the difference between the formal city of patricians and the informal city of the dispossessed. Avenida Paulista offered a luxurious alternative to the old city centre, with a focus on contemporary urban theories of sanitation, efficiency, and civility. The distance from the city centre and the design standards of the avenue gave the common areas of the avenue an exclusive character. As part of this model, Paul Villon, a French landscape architect who resided and worked in Rio de Janeiro, designed a small park, taking advantage of an existing wooded area located less than a kilometre from the western end of the Avenida Paulista. The collective facilities of the park, which included a restaurant on the south side of the avenue, officially opened on 30 March 1892, while the north side remained unoccupied for a couple of decades and offered a general outlook to the city. The park was named Parque Paulista (Paulista Park) but was often referred to as Parque Villon (Villon Park) in recognition of the designer's achievements.

Twenty years after its opening, Avenida Paulista became the main artery of the most affluent residential neighbourhood in São Paulo, with large villas in different architectural styles dispersed along the boulevard. Residences in Art Nouveau, neoclassical, Tuscan, and neo-colonial styles set along manicured avenues of trees emulated the urban life of the European Belle Époque (Sevcenko, 2000; Lima and Carvalho, 1997). Parque Villon was open to public use, but it remained private land until 1911, when Mayor Raimundo Duprat, who was also responsible for Joseph Antoine Bouvard's plan for the embellishment of the historic centre, decided to buy the whole property. Part of the plan was to redevelop the lot on the north side of the avenue as a place for the local elites to meet. He hired architect Ramos de Azevedo, the most prolific architect in São Paulo at the time, to design a large concrete structure with eclectic-style ornamentation on two terraces overlooking the city centre. The esplanade, finished in 1916 and named Belvedere do Trianon (Trianon Terrace), contained three pavilions and two pergolas on the upper part and a large luxury hall with restaurant and ballroom on the lower terrace. In 1918, British landscape architect Barry Parker redesigned the remaining south area of the park, which was remodelled and officially named Parque Trianon (Trianon Park). The project unified the urban blocks previously acquired by the city government in order to emphasize its public use, and to strengthen its spatial continuity with Avenida Paulista.

Figure 3.2. Trianon Terrace and ballroom as designed by Ramos de Azevedo. (*Photo*: Guilherme Gaensly. *Source*: Archive of Departamento de Patrimonio Histórico, São Paulo City Hall/DPH–PMSP)

Filling the Void: Ecstasy and Distress in the Paulistano Belle-Epoque

For almost 20 years, until the Great Depression affected North America and the capitalist world, Trianon Park and the terraces were one of the main urban spaces staging the cultural, social and political life that represented the frenetic ascendance of the commercial and industrial elites of São Paulo. It was also a significant example of how spatial and socio-cultural segregation went hand-in-hand in the expansion of city. Trianon Park was the point of departure for *corsos*, which were automobile parades, organized during the celebration of Carnival and enjoyed great popularity among the elites in the first decades of the twentieth century. This was a more disciplined and socially controlled alternative to popular balls and street festivities that took place in working-class neighbourhoods of the city (Sevcenko, 1992, pp. 104–106).

The restaurant and ballroom in the park were used for formal and informal events, from afternoon tea and ice cream to gala balls, conferences and political meetings (O Estado de São Paulo, 13 June 1916, cited in Moraes, 1995, p. 132). Among the frequent guests was a young group of artists and intellectuals gathered around the charismatic figures of writer and musician Mário de Andrade, writer Paulo Prado and poet Oswald de Andrade. Together they organized the 1922 Modern Art Week, which officially introduced the reception, debate, and production of modernism in Brazil.[3] The event, held in the ostentatious Municipal Theater, built downtown 11 years earlier, made its contribution to giving the city the appeal of cosmopolitanism that suited the social imaginary of the progressive groups among the affluent classes. This modernist group had an impressive pedigree. Some of the most influential artists and writers among them came from wealthy families and were in close contact with the early *avant-garde* movements in Europe, particularly surrealism and futurism. As they 'tropicalized' the artistic revolutionary energies of the time, they confronted the cultural conservatism of the coffee aristocracy, a gesture that was epitomized in the Brazilwood (1924) and Cannibalist (1928) manifestoes written by poet Oswald de Andrade. In these pieces, he proposed that artists in Brazil should metaphorically devour the inflow of foreign cultural models, and produce a national version of digested modernism. Much of the work produced in this period referred to national issues, however this was not a popular movement. Even though writer Mário de Andrade and composer Heitor Villa-Lobos were two main figures of the modernist movement deeply concerned with cultural democratization, their project did not embrace broad social, economic or political reforms. The elitist origin of the movement did not address the disjunctions between the circulation of European modernizing projects and the archaic political and social forms and popular cultural practices in Brazil.

The exclusive use of Trianon Terrace and of Avenida Paulista is a good example of the ambiguous relationship between urban renovation and public spaces in São Paulo. This privileged status went untouched until the mid-1920s. In the meantime, there was another less lofty European influence that leveraged political and social dissent in São Paulo: the organization of unions and labour demonstrations, which started with the massive immigration of Italian workers to the city. The nascent middle and working classes, including military officers, were growing more and more dissatisfied with the national political and economic life, and they organized uprisings in the city in 1924. Motivated by the Revolt of the Lieutenants that took place in Rio de Janeiro, protest groups built barricades in strategic streets and public spaces in downtown São Paulo as well as along Avenida Paulista to demonstrate against the rural and urban oligarchic groups that had tight control of the federal government. In response to these conflicts, the president sent troops from Rio de Janeiro, and held São Paulo under siege. They bombed strategic areas of the city causing hundreds of casualties and significant physical damage to buildings and streets. Avenida Paulista was still a residential area and it was physically spared from the attacks but the public life of the city in general felt the seismic waves of more serious changes to come.

Spatial Hiatus in a Booming City

After the 1924 uprisings were contained, urban development enjoyed a few more years of prosperity in a climate of optimism that was represented by the slogan 'São Paulo, the Fastest Growing City' in Brazil. However, the financial crisis in the New York Stock Exchange in 1929 profoundly affected the trade and the monoculture of coffee in the country with serious consequences for emerging metropolises like São Paulo. This event changed the social structure and the political life in the city, creating the opportunity for new leading groups and later the emergence of a sizeable urban middle class after coffee producers and traders declared bankruptcy. Many of the families who resided along Avenida Paulista lost their fortunes and had to sell their properties to the emerging elite of merchants and industrialists that grew in the shadow of the coffee economy.

The period between 1930 and 1945 consolidated the economic and industrial leadership of São Paulo in the country and was also a time of major transformation in the role of Avenida Paulista as a public space. Following the events of 1924, the avenue became again the stage for political demonstrations, which led to the Constitutional Revolution of 1932. This time the dissent was in the hands of the local emerging elites who fought to regain control of the national political scene, and also of the symbolic space of Avenida Paulista. Demonstrators opposed the regime that made Getúlio Vargas president after a coup against the election of Júlio Prestes in 1930.[4] The political and economic elites of São Paulo proposed the separation of the state of São Paulo from the Brazilian federation, capitalizing on the image that São Paulo was the locomotive in the process of modernization of a country they no longer controlled. This confrontation generated a civil insurrection in the city, which lasted three months until contained by federal troops.

Even though São Paulo did not achieve its separatist goals, an interim democratic regime was established with a new Constitution in 1934,[5] but it was short lived. Getúlio Vargas regained power in 1937 becoming a dictator and establishing a centralized regime known as *Estado Novo* (New State) that lasted until 1945. His second government was the result of an alliance between civil and military bureaucracy and the industrial elite to promote the industrialization of the country. Together they reinforced state technocracy and populist nationalism with considerable impact on the economic and urban life of Brazil, but little social advance (Fausto, 2001, pp. 366–367).[6] Like many other Latin American countries, their goal was to replace imports with national production and to build basic industrial infrastructure.

These policies were particularly beneficial to the industrialization of São Paulo and the concentration of wealth and new urban development around Avenida Paulista. The city was expanding quickly and it became the focus of intense urban planning efforts. The city wanted to convey an image of economic power and creative energy associated with the idea of progress (Lima and Carvalho, 1997), emulating the image of North American metropolises such as New York and Chicago. As a result, two important, complementary processes linking urban planning ideas and real estate development affected Avenida Paulista at the turn of the 1930s. Initially, engineer Anhaia Mello proposed zoning and building laws based on the example of New York City. Subsequently, engineers Prestes Maia and Ulhôa Cintra devised a master plan for São Paulo illustrated by vast panoramas of a metropolis made up of dense urban blocks, monumental buildings, and an efficient system of avenues. The scheme was based on Burnham's plan for Chicago among other examples.[7] In 1930 Prestes Maia, while the city's secretary of public works, developed the scheme into a plan known as *Plano de Avenidas* (Plan of Avenues).

Maia's master plan defined a succession of concentric rings and incorporated Avenida Paulista as a main artery connected by a series of coaxial avenues emanating from the historic centre of the city. This was the first (and also the last) time that morphology and aesthetics were part of a comprehensive reflection on the urban design and development of São Paulo. The main goal of this ideal plan was to create an overall model for the development of São Paulo based on expressways. The model also promoted urban decentralization, provision of low-cost, high-density housing, and motor vehicle circulation. Prestes Maia was appointed mayor in 1938 during Vargas's regime. He gradually implemented modified parts of the master plan, which promoted the construction of high-rise buildings based on setback and floor area ratio (FAR) legislation and on the expectation of mortgage policies.

Maia's mandate ended in 1945, without seeing his Plan of Avenues fully completed, and leaving the city with a deficient regulatory system of streets, transport, and infrastructure. With urban development in the hands of real estate interest, urban development privileged the circulation of cars over pedestrians, and efficiency over socialization. Consequently, public spaces became scarcer as the city grew without a system of collective open spaces to compensate for the increasing high-rise development. Moreover, after the 1940s the irregular concentration of urban investments strengthened the separation between wealthy central areas and impoverished peripheries with the concentration of public and private investments in the south-western areas of the city.

The imbalance between aggressive speculative activities by the real estate market, the limitation of Prestes Maia's plan, and the subsequent absence of regulatory plans together produced a new pattern in the city defined as *verticalização* (vertical growth), which is best exemplified by the boom of high-rise construction in São Paulo between the 1940s and 1960s (Somekh, 1997). The morphological transformation of Avenida Paulista from the pattern of large and isolated residential properties into the one of mixed-use high-rise buildings is a significant example of this development process. Typological changes came together with density changes, and massive buildings replaced the gaps left by large domestic gardens. These general changes eventually affected, as we will see below, the uses and the purpose of Trianon Terrace as a space of collective reference in the city.

The new generation of industrialists and investors that occupied Avenida Paulista in the 1930s and increasingly in the 1940s saw great opportunity for real estate development along the avenue. Avenida Paulista offered high symbolic and land value, given its unparalleled prestige, location, and infrastructure. The emerging elite wanted to invest in mixed uses along Avenida Paulista, and pressured legislators to change zoning and building regulations in order to demolish existing villas and to build residential and commercial highrises. In 1936, the city government approved legislation allowing commercial activities and the construction of residential buildings on Avenida Paulista for the first time. The process of vertical growth accentuated in the post-war period and, together with new urban regulations, Avenida Paulista attracted new residents and traditional commercial and financial activities previously located in the historic downtown.

The years between 1945 and 1964 represented an important period of democratization and industrialization in Brazil. The restructuring of capitalism after World War II together with the consolidation of architectural modernism in Brazil and President Juscelino Kubitschek's nation-building project changed the urban image of São Paulo. Kubitchek aspired to modernize Brazil in a very short period during his mandate. His national development plan, which followed the slogan '50 years in 5 years', culminated with the construction of Brasília as the new federal capital. São Paulo largely benefited from this plan both in terms of urban and economic growth and in social and cultural transformation.[8] The development of previous decades during the Vargas regime had given São Paulo favourable conditions for the reconnection of Brazil with the international economy. While agricultural and commercial activities had propelled urban and economic growth at the turn of the twentieth century, this time the connection happened with the boom of industrial and financial activities in São Paulo.

Avenida Paulista became the favoured place for the installation of national and multinational companies and banks in the early 1950s, while industrial production – headed by European and North-American car makers – grew in the western municipalities of the metropolitan region. During this period, São Paulo attracted massive migration from poorer areas of the country, and the population went from 2,662,786 in 1950 to 4,739,406 in 1960 (Sevcenko, 2000, p. 78). With industrial development came the creation of labour unions, which has had an enduring effect on the social and political life of the city and the country ever since. The concentration of social, economic and cultural capital towards the south-western areas from the historic downtown continued, causing even more separation between a wealthy centre and the ever-expanding impoverished peripheries. These events dramatically transformed the image and the social constituency of Avenida Paulista as an important symbolic place in the city. The new urban landscape of residential condominiums, office and commercial high-rise buildings, and cultural institutions[9] largely attracted the middle-class population.

Trianon Terrace did not survive the cultural and physical transformations that took place along the avenue. In the late 1940s, the park was losing its select users and the old ballroom was increasingly used for popular parties and balls and lost its exclusive appeal (Moraes, 1995, p. 63). The emergence of a mass cultural industry in Brazil played a significant role in this process. Popular bolero balls with radio celebrities and informal parties replaced the formal events at Trianon terrace. As a result, the cultural and social elites lost control over the use of the space, taken over by performances that they considered aesthetically questionable. These events, together with the traditional architecture of the park, conflicted with the lofty modernist aspirations that accompanied the increasing symbolic and market value of land along Avenida Paulista. Mayor Adhemar de Barros closed the ballroom in February 1951 and ordered the demolition of the whole ensemble leaving the site completely empty. Later that year, the area returned to the control of local elites as Ciccillo Mattarazzo, heir to one of the most affluent industrialists of São Paulo, sponsored the construction of temporary pavilions for the first International Art Biennial in São Paulo.[10] The city administration tried to develop the area as a cultural centre but the legal determination to maintain the view open towards the city as defined by Joaquim Eugênio de Lima, the original developer of Avenida Paulista, limited architectural proposals for the area.[11] The site previously occupied by Trianon Terrace remained empty until the late 1950s, when the directors of the Museum of Art of São Paulo saw a unique political opportunity to use the lot in order to relocate the museum from its temporary facilities in the historic centre to a permanent address on Avenida Paulista.

Reshaping the Void: The Transfer of MASP to Avenida Paulista

Trianon Terrace had been an important landmark to the affluent population of São Paulo until the 1950s. After this period, it became a more popular reference point in the city, thanks mostly to the expansion of the Museum of Art of São Paulo and to its transfer to the site on Avenida Paulista. The creation of the museum in 1947 was an important turning point in the social and cultural life of São Paulo. Assis Chateaubriand, a controversial and visionary press magnate, wanted to create a museum modelled on the Museum of Modern Art in New York City. MASP was housed for 19 years in the headquarters of his company *Empresas e Diários Associados* (Associated Daily Press) in the historic downtown. In 1946, he met the Italian art dealer and journalist Pietro Maria Bardi in Rio de Janeiro, where Pietro and his architect wife Lina Bo had come to supervise two shows and the sale of artworks from his gallery in Milan.[12] Shortly after, Chateaubriand hired Bardi to direct the museum and the couple moved to São Paulo in 1947. The Bardis created MASP based on the museological ideas fostered by ICOM (International Council of Museums), which was associated with UNESCO in the postwar period. These new trends stated that museums should go beyond their status as places of contemplation and engage in broader pedagogical and cultural functions, with potential to develop a more socially inclusive agenda. In the following decade, MASP became the epicentre of new artistic movements in São Paulo and expanded to three other floors of the building after several renovations to accommodate its increasing collection and activities.[13] The expansion of the museum required a new building and, in 1957, Lina Bo Bardi presented a proposal to use the lot once occupied by Trianon Terrace with an architectural design that respected the legislation and kept the plaza unobstructed.

The location for the new museum building is testimony of the Bardis' and Chateaubriand's political ability. Lina Bo Bardi suggested using the lot on Avenida Paulista after studying several possibilities around the city. Chateaubriand used his power in the press to propose an agreement with mayor Adhemar de Barros: if the city provided the site for the construction of the museum, Chateaubriand's newspapers, radio stations, and TV channel would support the mayor's gubernatorial campaign. Lina Bo Bardi had sketched an architectural proposal for the museum based on a study she developed a few years earlier for a museum in the coastal town of São Vicente. All parties reached an agreement and the project went ahead. The design proposal was the combination between the previous typology of Trianon Terrace, shaping the base of the museum as a semi-buried block containing auditoria, library, exhibition halls and a restaurant, and the Miesian transparent block for the permanent collection hanging from two long pre-stressed concrete piers – a solution that relied on the invaluable contribution of engineer José Carlos de Figueiredo Ferraz. Together the legal restrictions and modernist design principles, such as the use of simple structural logic elevating the building from the ground, created a powerful presence for the museum on Avenida Paulista. This unusual form defined a new image and a new space of collective reference in the city and helped project the image of the museum and the city abroad. Not only did Lina Bo Bardi's proposal preserve and reconceptualize the historic open space of Trianon Terrace, the design of the new structure created a prominent urban landmark of formal and chromatic contrast with the surrounding landscape, as the museum activities generated a new urban magnet in the cultural life of São Paulo.

Lina Bo Bardi's architectural gesture had significant political and cultural meaning. She recreated Trianon Terrace as a permanent open space to be used formally by the museum and also informally appropriated by the population of

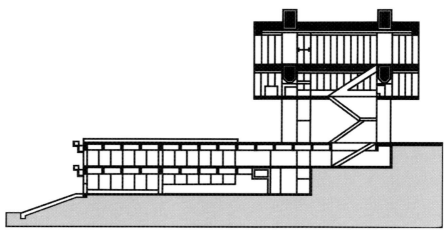

Figure 3.3. General section of MASP showing the lower volume with Trianon Terrace and the elevated volume suspended by concrete piers. (*Drawing*: Vera M. Pallamin, 2005, based on Lina Bo Bardi's and Figueiredo Ferraz's project. *Source*: Archives of the Instituto Pietro Maria e Lina Bo Bardi)

São Paulo. In one of her watercolours for the project, she imagined the terrace being used for formal and informal activities that went from art exhibitions to a large, colourful playground. The use of terrace as the mediation between the activities promoted by MASP and the city was one of the major innovations of the museum. As the Bardis expanded their experiments with the display of works in the temporary museum facilities in the 1940s, they stressed that the museum should abolish the traditional separation of the permanent collection in compartmentalized rooms. From a conceptual viewpoint, both the transparent volume containing the permanent collection and the plaza underneath should be as visually permeable as possible to its users and to the city. The plaza housed a few open-air exhibitions in the years following the museum opening, which included in 1969 a large exhibition of the works of Nelson Leirner, a Paulistano artist who was interested in participatory work, inviting the public to touch and move the sculptures. The strong presence of the terrace on Avenida Paulista was clearly defined by architect Aldo van Eyck, who wrote that 'the building is both there and not there giving back to the city the space that it ha[d] taken from it' (van Eyck, 1997).[14]

A New Urban Space is Born as Urban Democracy Retreats

Lina Bo Bardi conceived the museum on Avenida Paulista in the late 1950s, a time of great local and national enthusiasm, economic development, and political democratization in tune with the international changes occurring in the immediate aftermath of World War II. However, when the museum opened on 7 November 1968, the country was living a difficult moment in its recent social and political history. The climate of democratic life and industrial modernization became unstable in the early 1960s, when populist president Jânio Quadros renounced his position, leaving deep social and economic problems behind. João Goulart replaced him as a progressive leader, but did not have the same political and popular clout and was not able to contain the problems faced by the country. Protests began to take place in Brazilian cities including the largest of them on the streets of São Paulo, which gathered 700,000 workers in a general strike. This complex political situation was aggravated by economic recession, currency erosion, and high inflation, culminating in military intervention in March 1964 which coincided with similar events in Latin America during two of the most heated periods of the Cold War. The military coup in Brazil established a period of more than two decades of political authoritarianism, social control, and cultural censorship.

During the military regime in Brazil and particularly in São Paulo, local and federal governments concentrated investments in the development of large-scale urban infrastructure and state-controlled heavy industry. As this economic project took shape, the support for cultural and social welfare programmes for the growing urban population decreased. São Paulo concentrated the formation of large oligopolies that controlled the national economy in the following decades. Avenida Paulista largely benefited from these changes during the early 1970s and from the disinvestment in the historic centre during the two previous decades. In 1971, Mayor Figueiredo Ferraz, who had collaborated with Lina Bo Bardi in the design for MASP, promoted a large urban design project for the avenue known as *Nova Paulista* (New Paulista). This project created new sidewalks separating pedestrians from vehicular traffic and prioritized the increase of lanes for cars and mass transit, connecting several prosperous areas of São Paulo. Private developers tore down many of the original villas overnight in order to build corporate high-rise buildings on the highest point of the city.

As the venture proceeded, modernist highrises of the 1960s that followed a local, Brazilian character started to share space with the neutral curtain-wall structures of North American corporate architecture and central business districts. However, the short period of economic prosperity in the early 1970s known as *milagre brasileiro* (Brazilian miracle) did not survive the international oil crisis of 1973 and was replaced by profound economic, social, and political instability. Like Prestes Maia's earlier project for the city, the project envisioned by Mayor Figueiredo Ferraz for Avenida Paulista was not fully realized. However, the changes along the avenue were substantial, and helped consolidate the transfer of the main economic and cultural centre of the city from the historic downtown to Avenida Paulista.

In the meantime, the oppressive nature of the military regime substantially changed the collective and public life of Brazilian society with visible impact on the use of urban open spaces. Dissent was outlawed. Censorship and repression played a central role in the control of social and cultural expression in public spaces, in the media, and in universities until the late 1970s. Four years after the coup, in 1968 the military junta issued AI-5 (Institutional Decree, number 5), imposing strict control over citizens' individual freedom, effectively putting a straightjacket on social and political struggles in Brazil. This legislation gave military presidents ample control over the National Congress as well as over state and local representatives, including the power to appoint and to end the mandate of mayors and governors. During this period of repression and economic crisis, segregation became substantially stronger. The impoverishment of rural populations accentuated migration flows towards major metropolitan areas such as São Paulo, which had no systematic plan for the provision of jobs and housing to accommodate the expanding urban population.

The city of São Paulo concentrated investments in the remodelling and expansion of infrastructure and public services in consolidated urban areas, while the growing and impoverished peripheries remained underserved by public efforts at modernization. Subhuman living conditions increased in large cities, and shantytowns and tenements quickly expanded, reinforcing the spatial, social, and political differences in Brazilian metropolises. A large contingent of the population did not have access to citizenship rights, raising the level of discontent among those who opposed the military regime. The pervasive climate of economic depression and political oppression during the 1970s found increasing unrest in political movements organized by unions, left-wing intellectuals, and the progressive wing of the Catholic Church. These movements of resistance had great impact on the course of the authoritarian regime, which ended up lifting AI-5 in 1978. This political act initiated a long process of political amnesty and the end of official censorship. Another result was the strengthening of union movements, which organized several strikes in São Paulo between 1978 and 1979, and mobilized millions of workers locally and nationally in demonstrations in urban open spaces such as Avenida Paulista and through telecommunications media.[15]

In the meantime, the Museum of Art of São Paulo maintained its position of prestige in the cultural establishment of the city. The Bardis continued the cultural, pedagogical, and artistic mission of the museum after the death of its creator, Assis Chateaubriand, early in 1968, a few months before the opening of the new building. Lina Bo Bardi contributed to the museum as a curator in the late 1960s after her experience directing the Museum of Modern Art of Bahia in Salvador between 1960 and 1964, which was a key venue for the neo-*avant-garde* movements in that city. Lina Bo Bardi's sharp political approach to cultural production conflicted with the official views supported by the military regime. As a result of her increased participation in underground politics, she had to leave the country several times between 1969 and the early 1970s.

The museum entered a phase of ambiguity as Pietro Maria Bardi embedded it in the local cultural establishment and increased its audience by promoting popular concerts, classical recitals, and a film festival in addition to its regular activities of exhibitions, meetings, and art history courses. During this period the most important event to take place on the terrace beneath the museum was an exhibition organized with the collaboration of Lina Bo Bardi in 1972, which celebrated the fiftieth anniversary of the Modern Art Week. In a reverential – and at the same time provocative – gesture, she had a circus installed on Trianon Terrace. This risky gesture honoured the more progressive intellectuals who met in the Trianon restaurant to discuss the introduction of modernism in Brazil in the beginning of the twentieth century. Above all it honoured Piolin, a popular actor and clown whose work with *Circo Piolin* (Piolin Circus) had great influence on ideas about modernist theatre and the importance of popular cultural manifestations in São Paulo in the 1920s.

Trianon Terrace in Times of National Dissent and Local Manifestations

The 1980s marked the beginning of significant transformations along Avenida Paulista that are still in progress. The gradual demise of the military regime in Brazil left behind a profound economic and social crisis, and affected Avenida Paulista and the museum in different ways. Because the avenue occupies a central location in a city with strong symbolic appeal and visibility, it soon became an important forum of demonstrations and political dissent. Twenty years of authoritarian regime left many marks, especially the increased gap between the social and political groups included and excluded in the public life of the country and the daily life of Brazilian cities. Traditional forms of public use gradually started to share the open spaces of Avenida Paulista with protests against the military government. In 1984, the avenue concentrated several demonstrations of a widespread movement called *Diretas Já* (Elections Now). This movement started with a meeting gathering 1.3 million people at Praça da Sé (Cathedral Square) in the historic downtown in order to oppose the regime and to demand presidential elections; and later unfolded in a series of marches on the streets of São Paulo. Such protests contributed to the gradual transition of the military command into the reconstruction of democratic life in Brazil, which was marked by an indirect presidential election in 1985.

Since that time, the museum terrace and Avenida Paulista have become one of the main stages for dissent in the public and political life of the city, especially in the early 1990s. The most important event was the *Caras-Pintadas* (Painted Faces) in 1992, enacted by a massive and noisy crowd in support of the impeachment of Fernando Collor de Mello who, ironically, in 1989 had been the first elected president since the end of the dictatorship. Meanwhile, the museum and particularly the terrace gained great notoriety by becoming an important stage for the cultural and political life of São Paulo. The city, in agreement with the museum administration, rented out the terrace for several shows and fairs including a weekly antiques market. The population of São Paulo also started to use the plaza as a departure point for political demonstrations.

Figure 3.4. Avenida Paulista and the Museum of Art of São Paulo during a public demonstration. (*Photo*: Instituto Pietro Maria e Lina Bo Bardi)

The strong public feeling of the late 1980s and early 1990s against military authoritarianism introduced new voices in the political dissent that continues in the country to the present date. Old groups repositioned themselves as new constituencies appeared on the national political scene. One of the novelties was the increased support for left-wing parties by the Brazilian population, mostly with the ascendance of *Partido dos Trabalhadores* (Worker's Party) in different spheres of the political life of the country. The other novelty was the struggle to position Brazil in the new geopolitical configuration of globalization. Fernando Henrique Cardoso, representing the centre-left *Partido Social Democrático do Brazil* (Social Democratic Party of Brazil), initially dominated the political and economic life of the country by adopting neo-liberal policies dictated by global capitalism since the mid-1990s. However, this new strategy has not had long-term success in stabilizing the economy and in promoting substantial improvement in the social life of the country. Once again, the reorientation of the international political sphere had great impact in the country, in São Paulo, and along Avenida Paulista. Since 1989, the city has had elected mayors swinging from the far-left to the far-right, continuously shifting the concentration of public investments from social programmes to private development.

In the 1990s, the potential for construction on Paulista Avenue reached the maximum ratio determined by zoning laws and the local real estate market pressured residents to leave their lower-rent apartments in order to transform them into high-profit offices (Frugoli, 2000, p. 120). Later, private and public investments in the city became scarcer and they moved south towards new areas of development along the Pinheiros River. As a result of declining urban maintenance and land value, Avenida Paulista started to lose its single, privileged position of urban centrality and its prestige as a residential and commercial address. Developers recognized signs of physical deterioration and promoted the emigration of real estate investments from Avenida Paulista to other areas of the city. At the same time, the city started to finance the physical renovation of strategic locations, focusing on an international and corporate image of urban amenities as a way of attracting investors from the global economy.

In the last 10 years, Avenida Paulista has been among the three major areas of urban redevelopment in the city, which also include the historic downtown and the southern area of the Pinheiros River. In order to contain the flight of investment from Avenida Paulista, former Mayor Paulo Maluf, from the right-wing party *Partido da Frente Liberal* (Liberalist Front Party), organized a committee in 1995 with representatives of business and financial institutions located on the avenue. The

Figure 3.5. The use of Trianon Terrace for an antique's fair has marked the life of the plaza. (*Photo*: Zeuler R. Lima, 2004)

committee, which later became *Associação Paulista Viva* (Live Paulista Association), was created to promote projects for urban revitalization and gentrification in the area. Its goal has been to recover the symbolic, distinctive appeal that the avenue enjoyed until the 1980s, aspiring to look like the thriving commercial and financial centres of New York City. These efforts envision an exclusive urban environment, juxtaposing accelerated consumption, and images of high technology with the experience of increased urban violence and social segregation.

Between 2000 and 2004, a left-wing mayor from *Partido dos Trabalhadores* (Worker's Party), Marta Suplicy, replaced 8 years of mandate in the hands of the right-wing *Partido da Frente Liberal* (Liberal Front Party). Despite this, the economic groups located on Avenida Paulista still had considerable control over the changes in the collective and public spaces of the street. The administration changes to the centre-left party *Partido Social Democrático do Brazil* (Social Democratic Party of Brazil), which leans towards neoliberal policies, had not changed this orientation much when this article was completed in 2006. The search for urban excellence has continued to prevail over the complex spatial, economic, and social configuration of Avenida Paulista in the works that have been implemented. For example, the city created legal mechanisms prohibiting public demonstrations and enforcing labour inspections that result in the persecution of street vendors.

Figure 3.6. The Gay-Pride Parade 2003 along Avenida Paulista with gathering around Trianon Terrace. (*Photo*: Vera M. Pallamin, 2003)

In the last decade open collective spaces have become increasingly defensive and privatized, with surveillance systems and private security guards. Although Avenida Paulista remains a unique place of social diversity in the city – from street vendors to business people, from soccer celebrations to political demonstrations – it has been reshaped to respond to local metropolitan developments and struggles for an image of international and, now, global appeal. As a result, the reality of a symbol of public and cultural life between the 1960s and 1980s has surrendered to the image of a highly controlled and privatized domain.

As part of this scenario, abrupt transformations occurred in the relationship between the Museum of Art of São Paulo and the city in the 1990s and early 2000s. The recent economic and cultural changes taking place along Avenida Paulista are intimately connected to changes in the museum and in the terrace beneath the building, both in their cultural purpose and collective practices of appropriation. Pietro Maria Bardi resigned as the director of the museum in 1990, after 43 years dedicated to the institution. In 1995, the board of trustees named developer and architect Júlio Neves the new director. He has proposed drastic changes to the mission of the museum and to the spatial organization of the building. His remodelling, according to the official plan defined by the new administration, aimed at 'several problems regarding the conservation of the building as well as the promotion of museological improvements and the updating of technological equipment'.[16] MASP has also lost its unique prominence in the city's cultural circles, since it has become part of a wider group of private and public museums and cultural institutions, which include *Centro Cultural Itaú* (Itaú Bank Cultural Center), *FIESP* (São Paulo State Industrial Board) Gallery and *Centro Cultural São Paulo* (São Paulo Cultural Center), all located along the area of Avenida Paulista.

Budget problems from cutbacks in public funds, the concern with property and personal safety in a city that became more violent, the increasing commodification of art and culture, and the growth of tourist industry activities have also affected MASP, restricting public accessibility and contradicting the purpose of the stairs designed by Lina Bo Bardi to connect the plaza directly with the exhibition spaces of the museum above. For example, after decades of free access and public support, the museum started to charge admission fees and to raise funds from corporations in order to promote its activities. The museum also had to take care of the physical deterioration of the concrete structure and mechanical systems, which were restored by the new administration. The risk of overload on the concrete structure that sustains the terrace led the museum to prohibit its use by large audiences, coinciding with the desire for legal control of mass political and cultural events on Avenida Paulista as envisioned by *Associação Paulista Viva*.

Despite financial problems, layout changes and restrictions on public use, the museum and mostly the terrace beneath have not lost their symbolic place in the social imaginary of São Paulo and in the physical space of Avenida Paulista. On the contrary, the convergence between MASP and the avenue still enjoys a privileged position without equivalence in the urban life of the metropolis. The museum continues to provide public visibility, and recent restrictions have not stopped the collective appropriation of this important place and landmark. After almost 40 years, it continues to be the primary urban reference for different social groups claiming cultural and political dissent. These manifestations cover a wide spectrum ranging from protests and strikes to demonstrations for social recognition such as the annual gay parade, and from artistic performances to collective celebrations such as sport championships.

Final Considerations

This chapter has shown how the unstable cultural, political, and economic history of Brazil and São Paulo produced and continues to produce different social and physical spaces along Avenida Paulista, in particular the public space beneath the large concrete span of Museum of Art of São Paulo (MASP). The genealogy of the meaning of such spaces is intertwined with the genealogy of conflicts that have been at play in the development of the city, the growth of its population, and the changing values that are associated with this process. The consideration of conflicts over the production of urban space according to the definition of dissent proposed by Jacques Rancière is an important recourse to expand the understanding of history, public spaces, public culture, democracy and citizenship as shared and disputed dimensions of urban life.

The struggle over collective urban spaces and the representation of differences in them are indispensable elements for the advancement and spatialization of democracy. They introduce new political subjects and new rules into cultural and social life, and create possibilities for widening the exercise of citizenship from the abstract realm of the nation-state to the concrete realm of urban spaces. Political participation based on dissent calls for the continuous reinterpretation of formal, stable and universal definitions. At a time when significant modern values are challenged (including the historic notion of city), difference and its confrontations provide new references for the theory, planning, and design of architecture and urban spaces.

The open space Lina Bo Bardi incorporated in her design for the Museum of Art of São Paulo is a good example of how designers imagine and represent collective spaces, and how they are socially produced, confirmed or rejected by different social groups. Despite its vulnerability to change, the current uses and appropriations of the areas around the museum still fulfil Lina Bo Bardi's aspiration for Trianon Terrace as a space for the exercise of free thought and spontaneous social practices. After over a century of transformation this urban void continues to be reinvented, remaining one of the extraordinary symbolic centres of ordinary and public life in São Paulo.

Notes

1. Rancière proposes the term *mésentante*, which has been translated as 'disagreement' in English. However, we prefer the term dissent, which suggests a more uneasy meaning according to the Portuguese translation.

2. The immigrant population, which was largely of Italian origin in São Paulo in the turn of the twentieth century, strongly influenced the political life of the city. After the end of World War I, the city experienced a series of strikes that largely affected the agricultural export industry. Urban working classes carried out protests and demonstrations against poor living conditions, influenced by social movements occurring in Europe at the time. They claimed the prohibition of child labour under the age of 14, night labour for women, and the establishment of eight-hour work shifts. Federal laws proposing to 'combat anarchism' contained this wave of strikes in 1921 (Fausto, 2001, pp. 249, 287, 295).

3. The debate promoted by modernists about architecture in Brazil included issues about national identity, the renovation of formal and spatial languages based on technical transformations and building demands. Gregori Warchavichk designed two houses in São Paulo in the 1920s based on the architecture of Adolf Loos, which were responsible for the introduction of the principles of the Modern Movement in Brazil.

4. In honour of an army officer who participated in the 1930 coup, Trianon Park had its name officially changed to Tenente Siqueira Campos Park (Ocké, 2003, p. 46).

5. This constitution, inspired by the Weimar Republic, established free and mandatory public elementary education and legislation for national security and labour such as the regulation of benefits, under-age and female labour, and paid vacations and weekly breaks (Fausto, 2001, p. 352). This same year, the National Congress elected Getúlio Vargas President of the Republic.

6. Vargas was a controversial figure, and his appropriation of the public sphere was ambiguous. He supported centralizing material initiatives with large control over public opinion. At the same time, he stood behind the populist image of 'protector of the working class' (Fausto, 2001, p. 373) by investing in social services, and creating labour and rental laws as the components of an incipient welfare state. The contradiction between his dictatorial regime and the support of democracy during the World War II in the Western world led the political opposition to gradually remove him from power.

7. The other references to Maia's plan were the Ringstrasse of Vienna, Joseph Stübben's schemes for Dessau and Cologne in Germany, and Bartholomew's plan for Saint Louis, Missouri.

8. The radical political and urban project of Brasília is a good example not only of the pitfalls of modern functionalism, but also of the limitations in the conception of citizenship as a singular form of association in the sphere of the nation state. Studies of this metropolitan area have shown that the physical, spatial, and social reality of Brasília is very complex, and it reproduces the contradictions and profound social inequality common to large Brazilian metropolitan areas. This being so, the citizens of Brasília have repeatedly inverted, neutralized, and contested the meanings predicted in the original master plan designed by Lucio Costa and Oscar Niemeyer.

9. New architectural programmes and typologies started to appear on Avenida Paulista at the time. The most significant example of mixed-use buildings is Conjunto Nacional designed by architect David Libeskind in the mid-1950s. Two juxtaposed slab towers set above a raised base with an arcade system beneath that was accessible from the surrounding streets compose the ensemble.

10. Ciccillo Mattarazzo, the heir to a large industrial fortune, created the International Art Biennial of São Paulo. He also created the Museum of Modern Art in 1949 to compete with Assis Chateaubriand's Museum of Art of São Paulo. Mattarazzo's most important contribution to the city was the construction of Ibirapuera Park in 1954, which was designed by Oscar Niemeyer and landscape architect Burle-Marx in an area south of the most affluent residential areas of the city. All subsequent International Biennial took place in the Biennial Pavilion designed by Oscar Niemeyer in the park.

11. This was one of the reasons why the design for a cultural centre presented by architect Affonso Reidy from Rio de Janeiro, in 1952, was not built. His design proposed a semi-buried building on the slope towards the valley of Avenida Nove de Julho topped by an exhibition hall supported by a series of columns that obstructed the view of the city (Bonduki, 2000)

12. P.M. Bardi was highly influential in promoting the modernism in Italy in the 1930s and early 1940s.

13. Assis Chateaubriand and the Bardis also created *Habitat* magazine with the production support of Diários Associados in order to promote the museum's activities and to foster the debate about modernist aesthetics and artistic production in Brazil, Europe, and the United States.

14. Aldo van Eyck participated in the last Congrès International d'Architecture Moderne (CIAM), which investigated critical alternatives to the early modern *avant-gardes*. His critique was based on the emphasis on the social dimension of architecture. His concerns resonated with Lina Bo Bardi's work, and he became one of the main promoters of her architecture in Europe (Van Eyck, 1997).

15. These strikes were responsible for the emergence of Luís Inácio Lula da Silva as the leader of the steelworkers union in the industrial areas of the metropolitan region of São Paulo (ABCD), who eventually became president of Brazil in 2002.

16. For more information on the official renovation plans for the museum according to the new administration, see http://www.masp.art.br.

References

Andrade, Carlos R. Monteiro de (1998) *Barry Parker – Um arquiteto inglês na cidade de São Paulo. Tese de Doutorado*. São Paulo: FAUUSP.

Bardi, Lina Bo (1967) O novo Trianon 1957–1967. *O Mirante das Artes*, no 5, September/October.

Bardi, Lina Bo (1994) Video interview. São Paulo: Instituto Lina Bo e Pietro Maria Bardi (ILBPMB).

Bonduki, Nabil (ed.) (2000) *Affonso Eduardo Reidy*. Lisboa: Blau; São Paulo: Instituto Lina Bo e Pietro Maria Bardi.

Cadernos Cidade de São Paulo (1993) São Paulo: Itaú Cultural.

Fausto, Boris (2001) *História do Brasil*. São Paulo: Edusp.

Frugoli, Heitor Jr. (2000) Centralidade em São Paulo: trajetórias, conflitos e negociações na metrópole. São Paulo: Cortez/Edusp.

Lima, Solange Ferraz de and Carvalho, Vânia Carneiro de (1997) *Fotografia e Cidade: da razão urbana à lógica de consumo: álbuns da cidade de São Paulo, 1887–1954*. Campinas: Mercado de Letras; São Paulo: Fapesp. Coleção Fotografia: Texto e Imagem.

Lima de Toledo, Benedito (1987) *Álbum Iconográfico da Av. Paulista*. São Paulo: Ex Libris.

Maia, Francisco Prestes (1930) *Introdução ao estudo de um plano de avenidas para a cidade de São Paulo*. São Paulo: Melhoramentos.

MASP (1978) *Museu de Arte Assis Chateaubriand, Ano 30*. São Paulo: Secretaria da Cultura, Ciência e Tecnologia do Governo do Estado de São Paulo.

MASP (1982) *A Cultura Nacional e a Presença do MASP, Referências de P. M. Bardi*. São Paulo: Fiat do Brasil.

Moraes, Flavio M.B. de (1995) Estudo crítico e histórico da Avenida Paulista. Master's Thesis, Campinas, IFCH/UNICAMP.

Ocké, Marcella de Moraes (2003) Parque Trianon: história, análise e subsídios para o desenho. Master's Thesis, Universidade P. Mackenzie, São Paulo.

Rancière, Jacques (1996) O Dissenso, in Novaes, Adauto (org.) *A Crise da Razão*. São Paulo: Companhia das Letras, pp. 367–382.

Sevcenko, Nicolau (1992) *Orfeu extático na metrópole*. São Paulo: Companhia das Letras.

Sevcenko, Nicolau (2000) *Pindorama Revisitada*. São Paulo: Fundação Peirópolis.

Somekh, Nadia (1997) *A cidade vertical e o urbanismo modernizador*. São Paulo: Edusp.

Soufek, Antonio Jr. *et al.* (2002) *Avenida Paulista (A síntese da metrópole)*. São Paulo: Dialeto.

Van Eyck, Aldo (1997) Um dom superlativo, in *Museu de Arte de São Paulo*. Lisbon: Blau, no page numbers.

Chapter 4

A Memorable Public Space: The Plaza of the Central Station in Santiago de Chile

Rodrigo Vidal Rojas and Hans Fox Timmling

The Plaza of the Central Station in Santiago de Chile constitutes the western border of the original city, at the opposite end of the Plaza Baquedano. This latter space, 'over there where Santiago ends', as Benjamin Vicuña Mackenna declared, was what divided historic Santiago and the neighbourhoods of the more affluent socio-economic groups from the beginning of the twentieth century. 'At the other extreme', the Plaza of the Central Station has been the symbol for Santiago's major train station where southbound trips originate and where southern immigrants disembark in search of employment and in pursuit of dreams in the capital of Chile. Since 1903 the plaza has seen both formal and informal commerce, hotels and low-priced residencies, the market city and the violent city, poetry and death, culture and warehouses, and hope and failure. The historian and poet Roberto Merino describes it thus:

As far back as memory can take me, the peripheral areas that surround the Central Station have been infamous of neck fights and aggressive brawls, something typically associated with this popular and populous zone... A periodical from the beginning of the century describes this zone in the following way: 'Buildings on the verge of ruin, long and endless streets that are poorly maintained, humble sheds and huts, low class commerce and bars, barracks and warehouses'. It also mentions the presence of 'dirty squatter settlements full of brothels and cafes'. (Merino, 1997, p. 142)

It is from this plaza that the street known as the Chuchunco leads to a land subdivision of the same name. These two inspired the phrase, 'over there by Chuchunco' referring to something very far from the urban area, in a place that was demonstrably rural. The reason for this was that in the early twentieth century the Central Station Plaza marked the frontier between the historic centre and the rural surroundings. At that time, in an attempt to prevent the spread of epidemics and diseases, the Health Ministry had regulated the free movement of rural inhabitants, who were said to reside 'far over there, on the other side of the Central Station Plaza; over by Chuchunco'. This zone became the 'hygienic' frontier up to the mid-twentieth century.

Since then, the city has crossed that frontier and extended several kilometres westwards. The Central Station Plaza underwent gradual beautification. We see this in its new commercial gallerias, a pedestrian-friendly promenade, and new public, financial, and health-related services. Its public lighting and street furniture have also been upgraded. The space maintains its liveliness through diverse activities and functions, creating a vibrant social environment. One cannot help but see how this urban public space is the site for the enactment of all illusions and disillusions; of the social expression of discontent; of political vindication; of the pressures for survival; of the reflection of economic development; of the escape towards the south; and of an entrance to the metropolis. Definitely, the Plaza is one of the privileged places for the development of extraordinary events of significant urban impact.

Figure 4.1. The Grand Central Station Plaza in Santiago de Chile. This 1995 image shows areas that were reclaimed for pedestrian space.

Building the Central Station and its Public Space

In the nineteenth century, Chile witnessed a marked growth in mining production. This was one of the primary causes for the transformation of methods of production, communication and transportation. These changes required the development of infrastructure for transporting prime goods to industrial zones and from there to commercial centres. As a result, the first railway in Chile and in all of South America, Copiapó, was inaugurated in 1851. This evolution posed new challenges even from an architectural point of view. Railway stations were required to meet certain standards such as creating sheltered spaces where passengers could safely and comfortably board and disembark from the trains, while having appropriate conditions for the loading and unloading of goods. The first structure of the Central Station dates back to 1857.

The location of the Central Station on the western edge of Santiago gradually transformed the rural character of that region to one rich in commerce, residencies, and hotels. The station became the gateway to Santiago for those arriving from Valparaíso or the southern region of the country. It became the home for those who arrived in the city with nowhere else to go. Unfortunately, the activities in the area together with the socio-economic conditions of those who inhabited it resulted in its being one of the most disenfranchised sections of mid-nineteenth-century Santiago society. It was a place dominated by bars, saloons, and brothels, all inviting danger and poverty (Latcham, 1941, p. 344). Between 1870 and 1930, the changes were significant. Some are quantitative and can only be indirectly estimated due to the lack of statistics for the neighbourhood during that period. The most marked change is demographic, for rural inhabitants quickly populated the city. The urban population reached 27 per cent of the country's total in 1875, increased to 43 per cent by 1902. Between 1885 and 1895, Santiago's population grew by more than 30 per cent, and by 1907, increased a further 22 per cent. If these rates of increase are assumed for the neighbourhood, this would translate into an increase of 4,000 to 5,000 inhabitants, in addition to natural growth. Different testimonies provide evidence of the presence of migrants within and on the edge of the neighbourhood. Novelist Joaquín Edwards Bello writes:

Behind the Central Railway Station, named Alameda for its proximity to the entrance of that spacious avenue ... a sordid neighborhood has risen without municipal support. Its streets are dry and dusty during summer, and swampy and muddy during winter. One can guess that it is another new neighborhood, one of those that sprout like mushrooms in the cities of the Americas. Promiscuous, disgraced women walk the streets and are accompanied by dirty laborers to the nearby brothel, located on Maipú Street, across from the Alameda. (Edwards Bello, 1965)

The first rail tracks to be laid linked Santiago and the city of Rancagua, to the south. In this first phase of railway development, a bridge was constructed over the Maipo River, under the supervision of Henry Meiggs. In 1857, the Santiago locomotive successfully travelled the 16 kilometres that linked the capital to San Bernardo. At the same time, a network of 'bloody rails' (so called because of the use of horse traction) criss-crossed the city transporting passengers and commodities to and from the station. In addition, an extension to the railway, known as el Ramal de La Cañada, ran through the Alameda, from the Central Station to San Diego Street. In 1884, the corporation 'Trains of the State of Chile'

Figure 4.2. The third structure of Central Station in 1903. Note the sharing of the space by stagecoaches, trolleys, and automobiles. Also note the presence of pedestrians and urban gardens, at a time when planning for pedestrian activity was starting to pick up.

was created, the train reached the Alameda, and the second structure of the Central Station was built.

The principal access to the third and final structure of the Central Station was located close to the Alameda. This lost importance when a new access was designed on the east wing, towards Exposition Street. The new access was emphasized with structures similar to the first building along with a specially crafted space for the arrival of stagecoaches. Plaza Argentina was inaugurated in 1903 (Ortega and Oyarzún, 1980, p. 62) towards the Alameda, reinstating the importance of such access.

Plaza Argentina, which might be better described as a widening of the Alameda, was destined to attract large numbers of people and to serve as a parking lot for a significant number of motor vehicles. Thus, vehicular traffic was connected to the centre of the city by means of its most important railway. Other concurrent events stimulated the development of the Central Station and its Plaza. On 18 September 1857, 600 gas lamps illuminated Santiago for the first time. The same year the first Central Station structure and the entrance building were inaugurated. By 1888, 1362 gas lamps and 626 paraffin lamps illuminated the city, encouraging the use of Santiago's public spaces after dark. Also in 1857, the railroad company inaugurated a stagecoach service which linked the station to the centre of the city. In 1882, the electricity company lit the façade of its central building with electric lights, and

Figure 4.3. The transformation of the Central Station Plaza in 2005 into a commercial plaza accommodating pedestrians and semi-formal vendors.

Figure 4.4. Transformations that have taken place between 1980 and 2005: (1) the Ecuador and Alameda intersection is cleared to open a pedestrian-friendly walkway; (2) the *Planetarium* of the University of Santiago, Chile is constructed; (3) *Pabellón Forma* of the University is built; (4) a new access for the University of Santiago is built; (5) the *Mercado Persa Station* is created; (6) the Commercial Centre Central Station is built; (7) a new Central Station Plaza is designed and constructed; (8) a new access to the station is built, which required pushing the passenger loading platforms back 70 metres and creating a plaza for access that is directly connected to the exterior part of the Plaza.

soon public buildings and commercial stores were also lit by electricity. By 1883, the Plaza de Armas was lit electrically thanks to a newly installed generator on site. The ubiquitous electric light quickly found its way throughout the city. In 1884, the second Central Station structure was built. And with the electrification of the city came electric streetcars which were introduced on 2 September 1900. One of the streetcars' principal stations was located in the Central Station Plaza, whose third building opened in 1903.

From 1930 to the present day, the Central Station's urban form and urban culture have continued to evolve. This continuous process is characterized by the constant modernization of the city's infrastructure and its accelerating economic success, particularly in commerce and services. These developments, however, have negatively impacted the residential uses in the area. The area became home to rural immigrant groups who sought employment in the city. The impact of this produced great changes in the existing housing structure.

Not So Ordinary Spaces

Extraordinary events make their distinct mark on an urban space in such a way that does away with that space's ordinariness. But, conversely, it is also the uniqueness of the space that contributes to the realization of the events. If this were not the case, extraordinary events could take place at any ordinary venue throughout the city. In Santiago, there are only three spaces that host the majority of these extraordinary events in the city: the Italia o Baquedano Plaza; the Civic Barrio in the periphery of the Palacio de la Moneda; and the Central Station Plaza. What are the distinct characteristics of those apparently ordinary spaces that make them suitable to host extraordinary events? It is possible to distinguish five dimensions that make our Central Station so unique.

◆   A high volume of commuters that would provide a readily-available mass audience for whatever extraordinary event took place.

◆   High urban visibility: all three spaces are located in the city's main avenue, La Alameda.

◆   A large, easily accessible open public space suitable for congregation and dispersal.

◆   A symbolic value that is significant for all Chileans, either because of its components or its location. For example, Plaza Italia o Baquedano symbolizes the frontier between Santiago's high income class and the 'rest' of Chile; Civic Barrio is the location for the Casa de la Moneda (Presidential Palace), revered by all Chileans; and the Central Station Plaza, gateway to the south of Chile since 1857, is representative of the dreams and illusions of immigrants.

◆   High public and private vehicular flows, which imply larger numbers of observers for an extraordinary event. These flows potentially interrupt traffic, amplifying the impact of the event on the city.

Extraordinary Events

In this study, we will consider an event to be a meaningful act in public space. Its occurrence may be presumed, but is not necessarily part of the activities planned for in the design of a particular public space. The space may be suitable for the event, but it need not have been necessarily created anticipating its occurrence. We consider an event to be extraordinary if it satisfies three requirements:

1.   It seldom occurs.

2.   Its repercussions transcend the impacts on physical space to become the subject of public debate in the media.

3.   The actions involved in such an event do not correspond to events the urban space was actually created for. This last point refers, for example, to the student protests that take place in Central Station Plaza because of its close proximity to the University of Santiago, even though the space was not intended for that type of event.

This allows us to distinguish three types of extraordinary events:

◆   Politically oriented events, such as marches and gatherings of medium size and relative spontaneity. The larger of these unfold in secure, open, and accessible avenues, stadiums, and parks.

◆   Social protests, as in the case of university student protests.

◆   Celebrations, for example those related to the victory of a sports team.

To understand how a space like Central Station Plaza, which is apparently ordinary, gradually transforms yet retains characteristics of a privileged urban space suitable for extraordinary events, we introduce the concept of 'urban memory' (Barthes, 1997). A society's history as it is inscribed in its territory sets the foundation for the culture that develops in that particular society. Citizens' culture is translated into social relations, some of which are significant in that they transform into an active form of urban memory. They determine and provide 'specific ways' to interpret reality and set standards for social interaction for that particular historic society. We are talking about 'valued referents' that citizens incorporate into their conscience and express when the occasion calls for social interaction and the formation of collective identities and shared ideals. In the urban memory section, the concept is set as the foundation for 'understanding and accepting that which is foreign' (Wandenfels, 1997, p. 91).

Urban Memory and the Transformation of Space

Urban memory is the product of the historical accumulation of culturally transcendent images. These constitute the single most important element in building up the significance of the actual urban space; an urban space that we will not only come to understand as a structure for collective use, but also as a public means for expression. Additionally, it is culturally and socially significant in the construction of collective urban memories. As a means for transcendent expression, urban space plays a strategic role in the construction of urban images that are potentially significant in transforming into important referents of the cultural identity of a society.

Urban experience, social mobility, and architecture's symbolism directly affect the perceptions and emotions of individuals. They generate and transform identities through urban images that represent life *here and now*, which orient relationships between individuals and society in public space. The mediating role of the urban image originates in those elements that are consciously shared among all citizens. As Waldenfels (1997) asserts, the incorporation of the urban image

has to do with the relationship in respect to others. Our relationships to one another form shared values that we experience in urban space as accessibility to the originally inaccessible. These conditions finally help develop self-consciousness of that which is foreign and yet became to be constitutive of our selves. (Waldenfels, 1997, p. 39)

With this in mind, we must understand what role new signs, new functions, new land uses, and new spatial configurations can have in creating new associations and connections that will allow us to surpass that which is ordinary and repetitious and remember and preserve those extraordinary events that took place in the same spaces but in distinct time periods (Zecchi, 1978).

This is precisely what has been occurring in Central Station Plaza starting with the construction of the University of Santiago's Planetarium and the new main university students' entrance in 1981. This new access stimulated a large flow of commuters, reaching 10,000 daily. From Monday to Friday and March to December, commuters using buses and metro lines converged there. The university, since its founding in 1849, had not had a principal access to this plaza, not until 1981 that is. It transformed the function of the space, stimulated new activities, and gave the space a new significance.

The most significant transformation, nonetheless, resulted from the advent of unprecedented extraordinary events: student protests. The Plaza had experienced different public events not typically considered ordinary – rallies, celebrations, etc. (see figures 4.5 to 4.6). In addition, during the *coup d'état* of 1973, in which a military dictatorship headed by Augusto Pinochet overthrew the government of Salvador Allende, the Technical University of the State (as the University of Santiago was then called) was the site of violent confrontations, especially since the military knew that the University harboured supporters of Allende suspected of urban guerrilla warfare against Pinochet. In 1982, the student protests were initiated with the grand opening of the university's access to the Station Plaza. The first demonstrations were to denounce Pinochet's government. These protests were then organized annually until 1989, resulting in the loss of life for many students and commuters, and more than a few police casualties. From 1990 to the present, the Plaza has served as a site for protests, meetings, confrontations, and civil disobedience but with a new, more restricted objective: improving higher education. These events have not been exempt from anarchist interests.

Figure 4.5. Protests in the University of Santiago in Chile in 1995.

Urban Image and Inter-subjectivity

The use of urban space may facilitate cultural integration and the redefinition of meanings of urban images. Urban experience is instrumental in untangling the processes of consciousness. An urban image can definitely leave its mark in the identity of citizens. It constitutes a primary and permanent way to incorporate new values and visions. In this manner, the strangeness and 'otherness' of that which is foreign is incorporated into both the social and individual consciousness.

Urban design has always taken into account the quality of life in the city and the cultural integration that can be facilitated by urban public spaces, including the contribution of urban images to the cultural development of society. Early commentators such as Max Weber suggested that social actions become such if they possess shared significance between subjects throughout an urban space, which contributes to inter-subjectivity (Weber, 1958). The capacity for inter-subjectivity in a society consists of the possibility of creating urban images that are collective, socially communicable and culturally compatible (Lynch, 1960, p. 8). Many inter-subjectivities can evolve in one particular urban space, as the everyday experience of urban space results in a multitude of urban images.

From Lynch's point of view, the inter-subjectivity of citizens is anchored, temporarily, to some urban images in which we recognize signs and configurations that are shared by society. The Central Station Plaza, through its structure and function, is a sign that is interpreted as a platform for social progress and mobility (the arrival in the capital, place of dreams and new futures).

An urban image is considered transcendent and culturally significant when drama and spectacle coincide. Drama plays itself out when we understand perfectly the historical transcendence of the social and cultural contents involved, such as in the case of the social events related to the advent of democracy. Without a doubt, there are valuable urban memories in Santiago related to these social dramas that we would justly like to build upon to construct new cultural identities significant to contemporary society.

The Decoding of Signs in Everyday Experiences of Space

We are talking about the meaningful signs related to land use, activities, transportation, spatial structure, and architectural forms, along with elements of publicity and modern mass consumption. All these signs should be able to bridge people's perceptions to form inter-subjective meanings, known and accepted as representations of a new urban culture looking to provide socially accepted values and signs of a citizens' identity.

Everyday life results from a combination of situations, land uses, and both repetitive and more contingent actions. Order, disposition of signs, spatial configurations, and land uses certainly regulate the ways of using, understanding, and representing an urban space. The expressive means that give rise to the urban image are decoded by the citizens to later be re-coded. It is in this process of decoding and re-coding that one finds the possibility of experimenting and constructing new inter-subjective cultural identities in the city.

Figure 4.6. Public Celebration in 1901.

It is thus interesting to observe how the entrance to the Station is used for acts of distinct nature, despite being reserved primarily for motor vehicles. For example, figure 4.6 shows citizens paying homage to Almirante Barroso and to Brazilian marines, in 1901. It seems that, due to the strength of signs and a certain implicit social consensus, the different actors (demonstrators, pedestrians, merchants, motorists) accept each others' presence, and understand their rights to be in that place. Each one appears to inhabit by right a space that they also feel is collectively owned. It is possible to observe that the space is primarily reserved for vehicles, but without denying pedestrians. However, from the 1960s, with the introduction of vehicles capable of higher speeds (and more noise and pollution), we see efforts to separate automobiles and pedestrians.

The relevance of an inter-subjective urban image is because it places us at the centre of a network with other subjects, providing us with new cultural horizons of shared identities and a feeling of belonging in a society. The inter-subjectivity arises from the relationship between urban space and social actors which, through determined experiences and shared knowledge, construct new identity linkages and a sense of social belonging.

Significant Urban Memory

We should distinguish significant urban images from those which simply arise in the processes of consumption, production, and recreation, or with simple situations of random dwelling and spontaneity in public spaces. What is common and ordinary in urban space can be transformed into a significant place if citizens' interactions are geared towards the maintenance of relevant urban memories in the social and cultural history of a society. The urban image is then a field of experience which can strengthen social and cultural integration in a city. Extraordinary urban spaces are then spaces in which we have maintained important urban images which inspire subjective-affective sentiments among city residents against processes of social disintegration and media-lead vulgarization of culture. If urban images cannot catalyse inter-subjective values, we would not have any possibility of maintaining cultural identities to ensure a stable and significant relationship between individuals and society (Casey, 1998).

Kevin Lynch (1960) describes how a citizen interacts with his or her city through the urban image and the elements that act as its *markers*. We are referring to those signs and configurations that are given meaning by individuals when they go about, live, and perceive the city. It is impossible to understand the totality of an urban space, in its complex reality, if we do not overcome the *a priori* reductionism that urban images make of the complex urban world. *Markers* prepare us for mediation between the particular and the general, order and chaos, sign and symbol. An extraordinary urban space is that whose urban images contain its markers. These are markers recognized by people, thus socially shared and capable of unchaining sensibilities and sentiments of cultural integration in society.

Thus one of the principal objectives of contemporary urban design is to be able to transform an urban space perceived as *common and ordinary* into a highly expressive medium reflective of an *extraordinary* cultural environment. Following Neil Leach, we will suggest that the *inter-subjective projection of identity* between subjects and the signs of an extraordinary urban space allows the feelings and emotions of the human soul to motivate us to change into social subjects. In Leach's perspective, an urban environment has the potential to create urban images capable of transferring and constructing identity. Those images can change into paradigms of values necessary for a society to meet conditions for the development of a distinct urban culture (Leach, 1999, pp. 121–144).

Economic Development, Social Change and Public Space in Chile[1]

Previously, we stated that an urban image becomes transcendent and culturally significant when drama and spectacle coincide. We claimed that the Central Station Plaza is one of the privileged stages of this spectacle; one whose drama is constituted by the economic and social development of the country. It is the stage for and of political change; in other words, a public space slowly constituted from 1857 to the present day. In what follows, we highlight some of the central chapters of the social, economic, and political drama of the country, whose spectacle has had some memorable days in this public space.

From Economic Prosperity to the 1891 Revolution

Between 1830 and 1880, Chile experienced robust economic growth, stimulated by its entry into the world economy as an exporter of minerals and agricultural goods (copper and silver; wheat and flour) and as an importer of the manufactured products of the industrial revolution. The period of economic prosperity continued until the 1920s but was unable to mitigate the poverty that affected a considerable proportion of the population. As a reaction to this condition, the new social movements formed in the mining centres, ports and cities, with artisans and labourers as the protagonists and cooperatives as its organizations. Then, with the transformation of the unions into associations of resistance, the emergence of labour political parties, and the appearance of union leaders, a new era began, characterized by greater emphasis on ideology which radicalized the postulates of the movement.

By 1910, there were more than 400 cooperative organizations and a growing number of labour unions of employees of the metallurgy industry, the railways, and the typesetting companies, among others. The celebration of 1 May (as worldwide workers' day) grew in popularity. In 1907, more than 30,000 people congregated for that event in the streets of Santiago. After 1917, the labour unions enjoyed rapid growth, carrying out some 130 strikes between 1917 and 1920 throughout all of Chile. From 1920 university students assumed an active role, through the Federation of Students of the University of Chile (Federacion de Estudiantes de la Universidad de Chile, FECH), in supporting the struggles of the working class. At this time but more especially during the 1930s, the incorporation of the peasantry into the Chilean social movement began.

By the turn of the twentieth century, the 'social question' in Chile had been thrust into the public sphere, especially after bloody repressions of popular protests and denial of petitions became commonplace after 1850. At that point, the division between the opulence of the oligarchic class and the harsh conditions of the working classes had reached proportions unmatched in the history of republican Chile.[2] The popular struggles continued for some time, manifesting the dichotomies already present during the 1870s and 1880s in the way in which the different groups conceptualized the actions: 'labour strike' versus 'peon insurrection', or 'organized petition' versus 'spontaneous popular riot'. Like all dichotomies, this one is also arguable, although the empirical elements that support such characterizations are numerous during the last decade of the nineteenth century and the first years of the twentieth.

The Central Station, at the periphery of Santiago and migrants' point of entry to the city, was stage for the popular and labour movements of the era. This explains why the general strike of 1890 and the civil war of 1891 – in which the partisans of a parliamentary system of government clashed with President Jose Manuel Balmaceda and the partisans of a presidential system of government – took place in the Central Station. The strikers' movement of the winter of 1890 was the first generalized confrontation between the classes in various regions of the country, in which the workers joined ranks with great spontaneity and no little ingenuity. The civil war contributed to the relaxation of social discipline. At the Central Station, riots, insurrections, and the ransacking of public and private properties took place. Towards the end of the conflict – August and September of 1891 – the intensity of these actions reached their culmination due to the momentary void of power that was generated in some cities and regions. The elite felt, as at many times in the past, the dangerous presence of the 'lower masses' as a communication from the General Direction of Railways to the Minister of Interior reveals:

The events of September 29 of the current year prompted the 25 person contingency of Battalion 4, who were guarding Santiago's Central Station, to abandon their posts and their arms, of which the multitude that surrounded the station took possession. The people of the region, once they found themselves armed, assaulted the baggage office and destroyed or took control of one part of the packages that were found there. Later, they went on to take control of the waiting room and ticket office furniture. They entered the accounting office, and took some of the furniture and squandered all the rooms of loading guides, tickets, statistical documents, etc.[3]

From the First Strikes to Women's Suffrage

The first national strike, which Francisco Frias Valenzuela (1993, p. 402) called the *first manifestation of the modern social struggle*, was the Strike of Valparaiso in 1903. It was followed by the Strike of Meat in Santiago (1905), the Strike of Antofagasta (1906), and the Strike of Iquique (1907). In the last strike, 10,000 workers participated demanding improvements in their working conditions but were brutally assassinated, as had previously happened in the miners' massacre in Santa Maria de Iquique, among many others. These strikes were the impetus for the rapid rise of the Chilean labour movement. Its main leader, Luis Emilio Recabarren, founded the labour socialist party in 1912, renamed as the communist party in 1922 and becoming a member of the worldwide International Socialist. Also notable in that era is the gradual emergence of a middle class of Chile. The middle class and the working class were making great strides in the struggle for workplace rights, among which was a series of laws including instituting Sunday as a rest day, the 'law of the chairs' (the right of labourers to be seated while working), laws regulating workplace accidents, the right to workplace cribs, retirement and social provisions for state railway employees, all of which were put on the statute book between 1907 and 1918. These served as precursors to the women's struggle for formal recognition in the country's political life. Chilean women finally got to vote for the first time in the 1934 municipal elections, and in 1952 in the presidential elections.

All these events and accomplishments were made possible by the high turnout at protests and demonstrations and through the leadership of the steering unions and their political allies. The Central Station, the place of arrival for people from the south of Chile, was the *entrepôt* for those who travelled to the capital to add their voices to those social movements. The esplanade (the incipient plaza) in front of the main building was the place for innumerable demonstrations of all kinds. Diverse buildings, of dubious quality and function, sprung up around this esplanade to house hotels, working-class eateries, shelters, and all sorts of commercial activity.

From Benjamin Vicuña Mackenna to Industrialization

Between 1872 and 1930 the urbanized area of Santiago doubled and its population increased from 129,807 inhabitants in 1875 to 712,533 in 1930 (De Ramón, 1992, pp. 220–221). Because of this and the continuous introduction of new functions to the city, a plan to remodel the city was proposed, led by Benjamin Vicuña Mackenna, Chief of Public Works in Santiago between 1872 and 1875. According to Mackenna, Santiago was divided into two sections at the time. The first was characterized as the 'proper Santiago, the illustrious, opulent, Christian city', and the second, the city of slums that, for him, was nothing but 'an immense sewer of infection and vices, of crime and of stench, a true stable of death' (De Ramón, 1992, p. 225). The Central Station straddled both sections of the city, one where the displaced were at home, but also one where the more affluent classes used to arrive or leave the city. It is a place of contrasts, of social diversity, of encounters and confrontations. When the working class expresses its discontent or the middle class struggled for vindication, or the delinquents expressed their random rage, the Central Station was their venue. Meanwhile, the dominant class used the station to exit or enter Santiago.

And while people from the dominant class took advantage of the convenience of the train to leave Santiago, the rural population also used the train to enter the capital, where from the end of the nineteenth century industrialization had taken hold. In fact, in 1897 protectionist regulations aimed at stemming the influx of imports were brought in, which allowed industry to perfect its production processes and train its workers. By 1928, Chile had 200,000 men and 90,000 women working in industry. In 1939, President Pedro Aguirre Cerda created the Corporation for the Promotion of Production (CORFO) with the intention, among other things, of impelling industrialization and strengthening the role of the state in economic development, which has been recognized as the process of import substitution industrialization that occurred throughout Latin America.

According to data provided by Armando de Ramón, in 1930 there were 2,417 industrial establishments in Santiago, 28.13 per cent of the national total. In 1980, the industrial establishments of the capital represented 62 per cent of the total nationally. The opportunities afforded by the capital city, which boasted an image of prosperity and modernism, attracted tens of thousands of people from the rural areas. This migration led to the physical expansion of the city, especially of working-class neighbourhoods (both planned and improvised), increasing the demand for housing, infrastructure, transport, electricity, potable water, and other services that the city was unable to satisfy quickly. This caused a new wave of protests by labourers discontented with their working conditions – some more violent than those in the past. Again, the neighbourhood of the Central Station was the scene of these extraordinary events.

From the Military Dictatorship to the New Economic Prosperity

During the central chapters of the social, economic and political drama of the country, some of which unfolded in the Central Station Plaza, the Military Dictatorship (1973–1990) left indelible marks. From 1973 to 1981, the country endured 8 years when even the most minor opposition to the Pinochet regime was violently repressed. But starting in 1982, a weak process of political flexibility began in what is known as the *opening* (*apertura*) of the government. That year marked the beginning of what would be known as the *Period of Protests against the Military Regime*. These events, which took place in different urban settings and on distinct dates (privileging the more politically significant ones), resulted in the death, injury, torture, and imprisonment of countless citizens over a span of eight years. Labourers, high school teachers, women, college students, disgruntled adolescents and former militants of the Popular Unit of Salvador Allende were the main protagonists of these events.

In 1981, the Technical University of the State became the University of Santiago of Chile. Since then, it has become one of the main bastions of Chilean leftist politics. This had a high cost to the students: Mario Martinez, the student leader of the university, was assassinated in 1986; the student Carmen Gloria Quintana survived being burned, while the student Rodrigo Rojas was not so lucky and was incinerated and killed in 1984. With the advent of democracy in the 1990s, the bloody days were over. Nevertheless, with the passing of years, new movements and student protests have taken place. These do not have the characteristics of the earlier ones, and are motivated by other concerns: discontent with the precariousness of the state's financing for higher education. The methods of policing are less bloody, but the police clamped down on protestors with no less zeal than during the dictatorship. Though having different objectives, there is no doubt that the demonstrators feel they are heirs of Mario Martinez, Rodrigo Rojas, Carmen Gloria Quintana and many others in the battle for social and economic justice.

Physical Transformations, Permanence of the Events

The aforementioned historical account allows us to visualize how the Central Station Plaza has catalysed the formation of collective urban images, which are transmitted from generation to generation creating a corpus of inter-subjectivity. This allows the Plaza to maintain its role as a place for the collective expression of discontent and hope for better times despite the almost complete physical transformations of its urban space. Nevertheless, the Plaza has not lost its role as a privilege place for extraordinary events. We explain this privileged role in terms of the concept of *urban memory*. It is possible to distinguish at least seven parameters that explain this apparent contradiction:

• The character and intensity of economic and social activities transcend the physical transformations specific to each era.

• The residential population has maintained similar socio-economic characteristics over the span of a century, in spite of the general improvement in the levels of income of the national population.

• Central Station Plaza maintains its role as an urban node, both separating and connecting the central city with the new peripheral urban areas.

• The inauguration of the Metrotren service (suburban train) on 25 October 1990, consolidated the Central Station Plaza's position as the *door to the south of Chile,* since this service meets the needs of localities situated between Santiago and San Fernando, 50 miles south of the capital. In 1990, Metrotren offered six daily routes (three each way), which was increased to 94 daily trips (46 going one way and 48 the opposite way) by 2005.[4] The service has transformed Central Station into an inter-modal station, which connects the city with the nearby region.

• This connectivity increases the social and urban impact of whatever extraordinary event takes place in the vicinity of the station. In order to illustrate this, it suffices to note that, while in 1994 Metrotren transported 1,871,329 passengers, in 2005 it transported 7,583,667 passengers, with an average of 21,000 a day.

• The Central Station Plaza, the access to the University of Santiago of Chile, and the surrounding paths and public spaces constitute a juncture of easily accessible spaces, a great diversity of uses and activities, all of which favours multiple forms of social encounters and exchanges.

◆ Finally, the Central Station has persisted in time as a principal ordering element of the city's built environment and urban image.

Conclusion

It is precisely these events beyond the realm of the ordinary that in one way or another reconfigure urban space in a process of mutual interdependence. The characteristics of these spaces may or may not facilitate the realization of these events. For example, Rio de Janeiro's Sambodromo was built in response to the need to provide a proper spatial structure to host activities relating to the city's famous carnival parade. In contrast, the fencing off of urban public spaces by municipalities to prevent youths from patronizing spaces is quite common. But if they do occur, these events also induce transformations of urban space. These transformations reconfigure the semiotic blueprint of the spaces. The spaces that welcome extraordinary events are thus in-and-of themselves quite extraordinary.

Notes

1. Architect Lucía Ponticas Luna contributed to the writing of this section and the following ones.
2. On the 'social question' see Morris (1967); Vial Correa (1981–1986, vol. I, tomo II, pp. 495–551 and 745–782); Cruzat and Tironi (1987); Grez Toso (1997).
3. *Archivo Nacional, Fondo Ministerio del Interior* (en adelante *AN, FMI*), vol. 1679 (Comunicaciones con varias autoridades 1891), Dirección General de los Ferrocarriles del Estado, Chile, N° 1982, oficio al señor Ministro de Estado don Joaquín Walker M., Santiago, septiembre 1° de 1891, s.f.
4. www.efe.cl, accessed January 2006.

References

Barthes, Roland (1997) Semiology and the urban, in Leach, Neil (ed.) *Rethinking Architecture*. London: Spon, pp. 166–179.

Casey, Edward S. (1998) *The Fate of Place*. Berkeley, CA: University of California Press, pp. 285–330.

Cruzat, Ximena and Tironi, Ana (1987) El pensamiento frente a la cuestión social en Chile, in Berríos, Mario *et al.* (eds.) *El pensamiento en Chile 1830–1910*. Santiago: Nuestra América Ediciones, pp. 127–151.

De Ramón, Armando (1992) *Santiago de Chile (1541–1991): Historia de una sociedad urbana*. Madrid: Mapfre.

Edwards Bello, Joaquín (1965) *El roto*. Santiago: Ed. Universitaria.

Frías Valenzuela, Francisco (1993) *Manual de Historia de Chile: desde la Prehistoria hasta 1973*. Santiago: Zig-Zag.

Grez Toso, Sergio (1997) *La 'cuestión social' en Chile. Ideas y debates precursores (1804–1902)*. Santiago: Ediciones de la Dirección de Bibliotecas, Archivos y Museos.

Latcham, Ricardo (1941) *Estampas del nuevo extreme*. Santiago: Ed. Nacimiento.

Leach, Neil (ed.) (1999) *The Anaesthetics of Architecture*. Cambridge, MA: MIT Press.

Lynch, Kevin (1960) *The Image of the City*. Cambridge, MA: MIT Press.

Lynch, Kevin with Banerjee, Tridib and Southworth, Michael (eds.) (1990) *City Sense and City Design: Writings and Projects of Kevin Lynch*. Cambridge, MA: MIT Press.

Merino, Roberto (1997) *Santiago de Memoria*. Santiago: Planeta.

Morris, James O. (1967) *Las élites, los intelectuales y el consenso. Estudio de la cuestión social y el sistema de relaciones industriales en Chile*. Santiago: Editorial del Pacífico.

Pinto Vallejos, Julio (1982) 1890: un año de crisis en la sociedad del salitre. *Cuadernos de Historia*, no. 2, pp. 77–81

Ramírez Necochea, Hernán (1986) *Historia del movimiento obrero en Chile. Antecedentes. Siglo XIX*. Concepción: Ediciones LAR, pp. 293–312.

Ortega, Oscar and Oyarzún, José (1980) La Estación Central, Seminario de historia de la arquitectura, Departamento de Historia y Teoría de la Arquitectura y Urbanismo, Santiago: Universidad de Chile.

Vial Correa, Gonzalo Historia de Chile (1891–1973), Editorial Santillana del Pacífico, Santiago, 1981–1986, vol. I, tomo II, 495–551 and 745–782

Waldenfels, Bernhard (1997) *De Husserl a Derrida: Introducción a la Fenomenología*. Barcelona: Piados.

Weber, Max (1958) *The City*. New York, NY: Free Press.

Zecchi, S. (1978) *La Fenomenologia dopo Husserl nella Cultura Contemporanea*, Vol. 1. Florence: Loescher, pp. 20–35.

Zolezzi Velásquez, Mario (1988) La gran huelga de julio de 1890 en Tarapacá. *Camanchaca*, no. 7, pp. 8–10.

Chapter 5

Lima's Historic Centre: Old Places Shaping New Social Arrangements

Miriam Chion and Wiley Ludeña Urquizo

When one contemplates the Historic Centre of Lima, powerful images come to mind. Consider for example the fine wooden texture of the colonial buildings' balconies, some of which are in pristine condition, while others are on the brink of collapse. Or maybe the renovated Plaza Mayor with its beautifully landscaped gardens, sometimes traversed by people in horse-drawn carriages, sometimes the setting for international modern sculptures, and at others the meeting place for demonstrators demanding decent wages. And one cannot forget the major thoroughfares, which are at times clogged with cars and buses, or filled with vibrant street vendors selling shoes made in China, or by hundreds of thousands of pious Catholics joining the *Señor de los Milagros* procession. The Historic Centre brings together the multiple voices and diverse interests which make up Peruvian society.

The Historic Centre is located on the right bank of the Rímac River, 10 miles inland from the Pacific Ocean, at the centre of Metropolitan Lima, a city of 7 million inhabitants. Lima accounts for one-third of the country's population and more than half the national economy. The Historic Centre has provided a stage for the economic and social life of Peruvian society over many centuries, but the end of the twentieth century marked a major change in its spatial configuration. Its main economic functions contracted but despite its long trumpeted demise, it acquired a new urban vitality based on cultural activities and political demonstrations. Here, we propose to explain this renewed urban vitality through the analytical device of spatial capital. This allows us to observe the alignment between the major cultural-political events and the historic buildings and public places and their regional, national, and international networks.

Place and Spatial Capital

Central cities in many Latin American countries have experienced alternating periods of economic growth and decline. The dramatic growth of the informal economy and the abandonment of the central cities in the 1960s contrasts with recent efforts to rehabilitate them, attract new activities, and bring a new urban vitality. The case of the Historic Centre of Metropolitan Lima illustrates how the municipal government was able to restore the quality of the built environment, which in turn triggered major cultural and political events involving a diverse population and leading to considerable expansion of spatial capital. This is a dramatic case because of its unusually speedy implementation, whereby a run down and marginal place was swiftly refurbished to become one of attractive plazas and streets with lively public events.

Spatial capital is proposed here as a way to understand the social, political, and economic actions in a particular physical setting and the projection of those actions, locally, nationally, and internationally, through social and information networks. This framework is rooted in the traditions of urban social theory. More specifically, it draws on urban development and planning debates that were formulated and applied in many Latin American cities during the 1970s, when the dramatic urban growth invited scholarly attention to the relationship between the built environment and social and political processes.

Here, urban space needs to be understood as a place of intersection of multiple social relations that contribute to its social production. Castells defined urban space by its social practices and the social relationships that provide space with form, function, and social meaning (Castells, 1996, p. 152). Furthermore, urban places were defined by Castells as spatial units where collective consumption and the reproduction of labour power and social relations are grounded (Castells, 1979). Lefebvre, from a philosophical perspective, provided a useful explanation of the complexity of social space that accounts for activities, knowledge, and ideology as well as networks and pathways that allow production and reproduction to exist.

Social space is produced and reproduced in connection with the forces of production (and with the relations of production). And these forces, as they develop, are not taking over a pre-existing empty or neutral space, or a space determined solely by geography, climate, anthropology, or some other comparable consideration… Social space contains a great diversity of objects, both natural and social, including the networks and pathways which facilitate the exchange of material things and information. (Lefebvre, 1991, p. 77, originally published in 1974)

In 1974 Lefebvre (published in English 1991) was already discussing the networks contained in space. The intent was not only to go beyond the physical realm to the social realm, but also to go from the immediate boundaries to those defined by the networks of production and information. These networks connect localities to an increasingly diversified pool of capital, resources, and information across the world. At the beginning of the twenty-first century, the process of global economic restructuring and transformation in technology has made these networks even more essential in our understanding of urban space.

Discussion of urban space covers a wide terrain when entering the global and information economy domain. Ohmae (1996) argued that localities and nation states were losing definition to broad international networks of trade, information, and power. Sassen (1994) contributed greatly to our understanding of how global cities like New York, Tokyo, and London increased their financial power at the same time as economic and social polarization increased in their local spheres of influence. Amin and Thrift (1994) emphasized the social and institutional construction of places. They argued that 'globalization does not represent the end of territorial distinctions and distinctiveness, but an added set of influences on local economic identities and development capabilities'. Along these lines, Smith (2000) proposed 'the construction of alternative representations of local traditions in the same place, thus making room for conceptualization of the local as a site of contestations over meaning and power rather than a reservoir of unitary local subjectivity'.

Unlike these studies which emphasized the spatial integrity of the locality, Castells (1996) pushed the boundaries of local and global debates to argue that 'between ahistorical flows and irreducible identities of local communities, cities and regions disappear as socially meaningful places'. Castells did not argue that cities would cease to exist as an urban form but that the logic of urban space and development changed and this change required a thorough discussion of information flows and networks. He proposed the space of flows as a new spatial logic in the complex territorial development process whereby places do not define the flows, but the flows define the places.

The discussion of places and flows takes a different approach in Massey's work (1994, 2005). She formulated three propositions to understand space (Massey, 2005). First, space is always under construction; it is never finished or ended, like history it is always unfolding. Second, space is a multiplicity of stories and identities, full of conflicts. Third, space is a product of interrelations and intersections. She argues that 'the spatial can be seen as constructed out of a multiplicity of social relations across all spatial scales, from the global reach of finance and telecommunications, through the geography of the tentacles of national power, to the social relations within the town, the settlement, the household and the work place' (Massey, 1994).

She emphasizes the need to understand not only the content of the flows but also who has the power to generate those flows. She argued that places in the global context are defined not only on the basis of the mobility of the population but also on the power of the population in relation to the flows and movements. Some social groups initiate flows and movement, others receive them, and others are constrained by them.

Castells and Massey thus formulated different perspectives in which cities are not objects but processes; Castells proposed a 'space of flows' in a global context and Massey proposed 'articulated moments in the networks of social relations' at multiple scales. Based on these ongoing debates on the multi-dimensional, multi-scalar and temporal configurations of space, we propose the concept of spatial capital to link (1) the social, economic and political *actions and actors* that define the content of the place, (2) the *physical setting* in which actions take place, and (3) the projection of these actions through *networks*. When these dimensions are aligned spatial capital expands. Conversely, when the physical setting no longer meets the needs of the actors or when the networks disconnect from the actions, spatial capital shrinks.

An understanding of these elements can highlight the tensions among actors and their interplay to appropriate the built environment, the development potential of a particular place in a broad regional and international context, and the reach of political and economic powers across places and across administrative and geographical boundaries. In Lima's Historic Centre, these actions have been centred on cultural-political events in a physical setting that carries much historic value and symbolic power. These actions create a powerful central node that projects a strong sense of identity and citizen participation at local, national, and international levels. These actions were reported through multiple conduits of media and institutional networks, engaging a very disperse population and replicating these actions in other places.

Power and Place across History

Over five centuries, the role of Lima's Historic Centre was defined and redefined as a centre of the region, the viceroyalty, and the country. This history is imprinted on the Historic Centre's buildings and streets, people and activities, and on its relationships with the rest of the region, country, and world, all of which shaped the expansion and contraction of its spatial capital.

During the first years of the Spanish conquest, the struggle between the Incas and the Spanish for political control of the country led to the creation of a new capital city in Lima. Around 1460, the *Señorio de Taulichusco*, Taulichusco's Dominion, had jurisdiction over many small villages spread throughout the three coastal valleys of Chillón, Rímac and Lurín and organized the provision of water, religious activities, and military forces. The major buildings and activities at that time defined the place that remains the centre of the city today. The Puma-Inti *huaca* or temple, dedicated to the sacerdotal caste and a place for the collection of religious offerings, became later the location of Lima's cathedral. The main *cancha* or courtyard for llamas became the site of the Spanish town hall and today's city hall. When the Spaniards arrived to establish Lima as the capital of the viceroyalty of Peru, the territory that today occupies the metropolitan area had more than 150,000 inhabitants. In contrast, by 1613 Lima was concentrated within 316 hectares and had a population of 25,000 people. While Lima was founded in accordance with the rules of Western centrality and Catholicism, it was also shaped by the original rationale and spatial logic of the Inca place (figure 5.1).

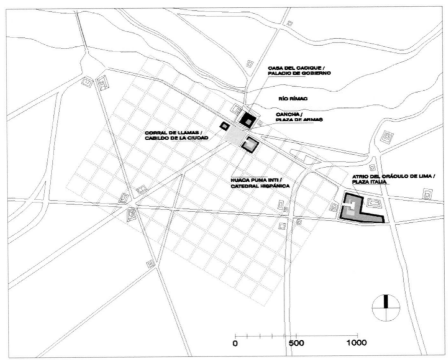

ORIGINS OF THE CENTER OF LIMA (1534):
TAURICHUSCO'S CENTER V.S. THE SPANISH CENTER
Adapted from Juan Günther Doering & Guillermo Lohmann Villena. 1992. "Lima". Madrid: Editorial MAPFRE S.A.
CAD: Marcos Aquino Casabona. 2005

Figure 5.1. Native city versus Hispanic city. Pizarro's downtown versus Taulichusco's downtown. (Graphic editing over Juan Günther and Guillermo Lohmann's schemes, 1992. *Concept*: Miriam Chion and Wiley Ludeña Urquizo. *CAD Drafting*: Marcos Aquino and Carlos Yagui)

After Independence and during the formation of the republic, the competing powers of the central government, the Catholic Church, and the business sector fought to secure the secular organization of society, a struggle that is still associated today with Plaza Mayor. In the nineteenth century, when the country was shifting from colonial to republican governance, an emerging oligarchy was attempting to establish itself as a major political player, based on agricultural and mining exports (Ludeña Urquizo, 1996). For this group, the city needed a secluded residential space and a recognizable financial and commercial core to control growing commercial activities and symbolize the new economic power. This led to the consolidation of the new urban form: a radial structure in which a residential periphery was connected to the new financial and commercial core.

It was not until midway through the twentieth century that the Historic Centre experienced its largest investment in infrastructure and real estate, the highest concentration of national and international financial institutions and of commercial activities in the city and the country at large. However, this phase did not endure; as soon as the peak of investment was reached the reversal began. The country's engagement in capitalism was more through consumption than production. The centralized resources in Lima triggered a huge wave of migration to consume imported industrial goods.[1] Then, when the new immigrants were confronted by the lack of employment opportunities, they turned to the informal sector for a living, which overwhelmed the Historic Centre. The immigrants invaded the centre, setting up informal homes and businesses at the same time as the exodus of the main commercial and financial businesses and major government and education institutions began. This heralded a new popular, *mestizo*[2] Andean centre that acquired increased social and cultural meaning for the immigrant populations of Lima's squatter settlements. The Historic Centre also acquired a new urban dynamic, a new face, and a new identity, all still evident today in the images of modern high-rise buildings against a backdrop characterized by vendors pushing carts along the streets.

The intensity of these new uses, the abandonment of the Historic Centre by central government and the scarcity of local government resources resulted in a serious deterioration of the environment, buildings, and streets. The exponential expansion of the informal centre peaked in the late 1980s. It was at this time that the city and the country suffered high levels of political violence and economic crisis, which virtually paralysed the Historic Centre (INEI, 1997; Iguíñiz, 1996; Crabtree and Thomas, 1998).[3] Peru's tense relations with international financial institutions, the result of President Alan García's decision to limit the foreign debt repayments, only aggravated the situation. In addition, terrorist actions by Sendero Luminoso and clashes with the military resulted in more than 70,000 deaths countrywide and major destruction of infrastructure. Random bomb attacks, shootings, and seizure of institutional buildings led the general public to avoid public places. By the end of the 1980s, streets were blocked and plazas were closed to public access.

These conditions changed dramatically in the mid 1990s. The decrease in politically motivated violence, the implementation of drastic economic policies, and the privatization of major state companies allowed Peru to experience robust economic growth as well as increases in international trade and foreign investment.[4] However, this dramatic growth did not necessarily translate into an improvement in the quality of life for the masses. The period was characterized by Lima's aspirations to become a global city, a task to be accomplished through President Fujimori's neo-liberal euphoria. When President Fujimori left the country in 2000, much political and economic stability was secured by interim President Paniagua and his elected successor President Toledo. By 2006, the collapse of the neo-liberal economic agenda in the country had pushed the voters to the left, a trend that was already taking place in many other South American countries. In the presidential elections, former president Alan García and his left-of-centre political agenda, defeated nationalist Ollanta Humala, the radical left candidate, by a very small margin as well as Lourdes Flores, the candidate of the right, by a very large margin. These historic events have conditioned the spatial capital of the Historic Centre today. From the time of the Inca Empire through the Spanish colonial times to the 2006 presidential elections, the major events in the life of the region, the city and the country were played out in the same place: the Historic Centre.

The Flight of Financial and Commercial Services from the Historic Centre

The recent changes in the spatial capital of the Historic Centre emerged as its economy underwent major changes. Its financial and retail sectors contracted dramatically as the city and the country became more bound up in new global networks of production and consumption.

Once the national centre of financial activity, the deterioration of the Historic Centre during the 1970s and 1980s, together with infrastructure and transport problems, made it very difficult for professionals in all sectors, but especially financial services, to conduct their business there. Thus many financial institutions relocated elsewhere, first in a scattered pattern throughout the more modern commercial and upscale neighbourhoods. Later, during the 1990s, as the country experienced a rapid growth in capital and international trade business, financial activities consolidated in a new financial district in San Isidro, a wealthy residential neighbourhood.[5] The few financial services that remained in the Historic Centre were primarily those direct services catering for local businesses and households. The new financial district exhibited the spatial impact of the expansion of global financial networks in Metropolitan Lima. Information-based activities operating as a network of services required a nodal site that could provide the appropriate office accommodation equipped with modern technology and homes for a professional talent pool; these were not to be found in the Historic Centre (Sassen, 1994; Hall and Castells, 1994; Chion, 2000).

The spatial capital of the Historic Centre capable of supporting financial and information-based activities thus suffered dramatic contraction. The prominent actors, such as banks and other financial institutions, took their activities elsewhere because the built environment lacked the appropriate form, function, and symbolic power necessary for their operations. Further, the Historic Centre could not provide the infrastructure or cater for the professional clusters needed to support essential metropolitan and international networks.

Relocation of Merchants from the Streets to Markets and Malls

The conquest of the Historic Centre by street vendors, and their subsequent removal were dramatic events in the life of the city. The population growth of Metropolitan Lima,[6] not matched by a parallel expansion of urban industrial activities, fuelled the emergence of the informal sector, in which street vending was a major sector (De Soto, 1986).

In the Historic Centre, the number of street vendors with permanent stalls increased from approximately 7,400 in 1976 to 9,500 in 1985 and to 17,000 in 1995 (Chávez et al., 1998; Instituto Libertad y Democracia, 1985). These stalls ranged in sophistication from a mere plastic surface on the sidewalk (on which the merchandise was displayed), to fixed wooden and metal structures that covered a large sidewalk area or traffic lane. By the late 1990s, approximately 20,000 street vendors occupied nearly 300 street blocks in the Historic Centre (figure 5.2). This explosive growth precipitated the deterioration of the existing build-ings and transport system, the closure of certain streets, and the high levels of pollution.

During the late 1970s and 1980s, most municipal governments avoided the issue of street vending or had very marginal success dealing with it. In 1995, Mayor Alberto Andrade accomplished the first successful attempt to regularize and relocate street vendors into various newly created retail centres. This success relied on the development of alternative relationships between informal businesses and government institutions, the revision of economic policies, and the reduction of red tape (Távara, 1994; De Soto, 1986). The relocation of street vendors allowed streets and plazas to be remodelled, improvement in transport services, and the

Figure 5.2. Ambulatory commerce in Jirón Lampa, 1996. (*Source*: Andrade, 1996)

regeneration of vacant lots and buildings, which resulted in increased property values. These efforts were deemed important and necessary in spite of the loss of income and the hardship born by those vendors unable to relocate to the designated areas.

Beyond the new retail centres, the relocation operation was facilitated by the availability of vacant commercial space in the vicinity and the financial resources of some of the vendors to start their own establishments. This relocation was also linked to broader retail and consumption patterns at the regional and international scale (Urry, 1995). Greater access to low-cost imported goods, primarily from China, and the development of shopping malls as the standard retail outlet effectively cut through the street vendors' price advantage. Major shopping malls were developed even in what were considered marginal neighbourhoods.[7]

Street vendors were important actors, who performed key retail activities, and developed strong links at the metropolitan level with most of the peripheral neighbourhoods. However, the built environment was never adjusted to accommodate these actors and activities. The form and function of the buildings and streets did not allow the proper functioning of street retail activities and the flow of workers and customers. In addition, the new international retail networks and consumption patterns demanded a different spatial typology that could handle a larger and more systematic exchange of goods at the international scale. Here, the international networks of consumption overwhelmed the metropolitan networks of consumption.

At the dawn of the twenty-first century, the Historic Centre retains a variety of business services, retail stores, churches, and schools. However, it no longer functions as the main economic engine of the city or the country as it did in previous eras. Its role as a financial centre and as the largest and most diverse informal market has been diluted. The old spatial capital of the Historic Centre, built around financial and retail activities, vanished as these activities could no longer function in such built environment nor develop regional or international networks.

Urban Stage: The Changing Face and Function of the Historic Centre

The recovery of the Historic Centre is not only essential for the neighbours of Lima and Peru, but also for all Ibero-America... In the context of the expansion and consolidation of the world market and globalization, cities are placed in a process of preservation and revaluation of their historical legacy, attempting to reinforce their own roots, their own identity... In the era of the world market and globalization, the reaffirmation of our history, our identity, the development of unity in diversity, become essential... The challenge that mayors face is to link tradition and modernity, past and present, and it is the challenge of our countries to integrate ourselves into the world knowing who we are, where we come from, and where we are going. (Mayor Alberto Andrade at the Summit of Latin American Mayors, November 1997 – authors' translation)

In the mid 1990s, the Historic Centre was becoming a stage for local history and culture with a renewed urban vitality. The municipal government had a comprehensive urban agenda to pursue this path in line with broader national and international trends. In addition to the relocation of street vendors, it implemented two other strategies: public space improvement and historic preservation, and sponsorship cultural activities.

The municipal government was an essential player in the overall urban transformation but it was not alone. The proliferation of cultural activities at the Historic Centre was also sustained by other groups, such as non-governmental organizations, artists, activists, and neighbourhood groups. The synergy between the municipal rehabilitation programme and the proliferation of cultural activities pushed the physical transformation of the Historic Centre from a mere cosmetic facelift to a major spatial and structural transformation that revealed accumulated spatial capital over time.

Public Space Improvements and Historic Preservation

The rehabilitation of the Historic Centre was not a new policy on the agenda of municipal governments; public space improvement had been a traditional populist practice in Peru. But it was only at the end of the twentieth century that municipal efforts went beyond superficial repairs, were sustained beyond the planning stage, and went on to successful implementation. Mayor Andrade's wide popular support and access to international funding made possible major public space improvements, which were initiated at a brisker pace and on a much wider scale than previous efforts.

Rehabilitation of plazas with the greatest symbolic value for the city and the nation was the first task to be accomplished. Among these was Plaza Mayor, the most important public space since the time of the Taulichusco's Dominion; Plaza San Martin, a key venue for political gatherings; Parque Universitario; Plaza Francia; and Plaza Italia (figures 5.3 and 5.4). This was followed by the remodelling and historic preservation of streets and major thoroughfares, such as Pasaje Santa Rosa and Pasaje Escribanos. The work encouraged the establishment of upscale cafes and bookstores catering for an upper middle-class clientele. Plazoletas Santo Domingo and San Agustín underwent similar improvements. The streets providing access to the main plazas were completely remodelled with new street furniture, including decorative lamp posts, benches, waste bins, and banners, among the first of which were Jirón Lampa, Avenida Abancay, Jirón Cusco, and Jirón Camaná.

One of the most successful historic preservation efforts was the 'Adopt a Balcony' programme, designed to save from destruction one of the key features of Lima's traditional colonial architecture, namely, the enclosed wooden balcony. Its design and construction have been transformed over the centuries, but the basic

Figure 5.3. Plaza Mayor. (*Photo*: Wiley Ludeña Urquizo, 1997)

Figure 5.4. Parque Universitario. (*Photo*: Wiley Ludeña Urquizo, 1997)

elements retain a strong *mudéjar*[8] influence. These are carved friezes supporting lattice windows, which are crowned with balustrades and cornices. By 1997, many of the 200 balconies in risk of collapse were rehabilitated (figure 5.5).

Improvements were specifically geared to the use of public spaces by local

Figure 5.5. Balconies of Osambela Palace. (*Photo*: Wiley Ludeña Urquizo, 1999)

and foreign visitors. This was exemplified by the work at the Central Market, the oldest grocery and produce market in Lima, and the adjacent Chinatown, both of which underwent dramatic transformation including the relocation of street vendors, reopening streets to public transit and taxicabs, restoration of historic buildings and the Chinatown gateway. These improvements encouraged many visitors and boosted business for many Chinese restaurants and retail stores. The restoration of the Rímac river banks in conjunction with the development of the Alameda Chabuca Granda, a grand boulevard, was another project to address environmental improvement issues and attract visitors. Even though the project was not fully completed, the boulevard became one of the most visited places in the Historic Centre.

Beyond the municipal efforts, financial institutions and other businesses sponsored historic preservation projects for churches and monasteries, such as Santo Domingo, La Merced, San Francisco, and San Pedro. They also sponsored the rehabilitation of *casonas*, or old residences, for use as office, cultural, and retail activities, many of which were converted into exclusive galleries and restaurants. In a few cases *casonas* were restored for residential use, as in the case of Quinta Heeren, a multi-family complex, and Casa de Berckemeyer, a single-family home, both built in the nineteenth century. The former was a private-public effort to provide a combination of middle- and low-income housing. The latter was the mayor's residence and therefore rehabilitation with a strong political statement in support of the newfound urban vitality of the Historic Centre, since no mayor had lived in the Historic Centre for decades.

The municipal programme went beyond cleaner streets and buildings. It also renamed places to emphasize the importance of the Historic Centre as a source of local identity. The downtown area that had been known simply as *Centro* was renamed *Centro Histórico*, or Historic Centre; the Plaza de Armas, was renamed Plaza Mayor; Capón became *Barrio Chino*, or Chinatown.

Cultural Activities

The rehabilitation and preservation of the Historic Centre, which took place when the country was recovering from years of political violence, was central to the development of cultural activities. During the late 1980s and early 1990s, streets and plazas were closed, and curfews imposed after 10 pm. Those were the times of night without night life, and streets without street life. This situation changed dramatically in the mid 1990s, when the attainment of political stability in the country fuelled a social euphoria in the use of public spaces, with people rushing to the streets to use and appropriate them. The Historic Centre was at the heart of this social euphoria.

The municipal government took the lead in the promotion of traditional and modern art forms. This mix was very attractive to a broad spectrum of people pursuing a newly reclaimed identity based on local traditions but perceived in terms of contemporary values and international trends. The municipal government sponsored tours of historic monuments and plazas, library workshops, concerts, festivals, theatre performances, and conferences. Free performances of traditional music and dance from the various regions of Peru were very popular. Many of these events were replicated at theatres and plazas in various neighbourhoods highlighting the connection of the Historic Centre with the rest of the city so making it a major metropolitan cultural node. Major national events included the First Festival of Music, the First National Fine Arts Exhibit, and the First Festival of Art and Electronics, all of which involved high profile artists. The international dimension of these efforts was exemplified by the First Ibero-American Fine Arts Biennial Exhibition that took place in 1997, 'dedicated to the study of cultural plurality expressed in new art concepts as well as the collapse of traditional concepts in what is understood as national art', as stated in the promotional brochure for the event. Paintings, sculptures, and installations by various artists from nine Latin American countries and Spain displayed using rehabilitated historic buildings and plazas as galleries and stages for the exhibitions (figure 5.6).

Beyond the municipal projects, public spaces were also brought to life through

Figure 5.6. Ibero-American Fine Arts Biennial Exhibition, Pasaje Olaya. (*Photo*: David Gonzales Zamalloa, 1997)

more casual, neighbourhood-based and small-scale activities. Food and crafts fairs were held in a number of plazas. Several cultural galleries were opened by neighbours and former street vendors to host exhibits, conferences, and music festivals. An example of a major informal initiative was the coming together of the former book and magazine street vendors in Jirón Quilca and Plaza Francia, which made the area an alternative cultural node of the city. This establishment of this node was catalysed by the Cultural Centre El Averno, led by the painter Herbert Rodríguez, where a counterculture movement was established. The centre hosted a variety of events from scholarly conferences, to art exhibits, to rock festivals. It also stimulated one of the most intense forms of public expression: the politically charged Averno style murals.

The Historic Centre, as a place with powerful cultural and historical symbolic references, provided a counterbalance to the strong drive in favour of global consumption expressed in the proliferation of shopping malls, fast food outlets, and gated-residential communities. During previous decades, the cost of access to international media had been prohibitive and was restricted to the local elite, but by the mid 1990s they had declined and enabled the creation of a very wide consumer base. This had the unintended effect of lowering the social status of international media products while boosting the value of local products featuring elements of traditional culture.

A Stage of Social Theatre

While the physical improvements involving the relocation of street vendors led to the displacement of a segment of the low-income population, and were designed to attract a higher income population, this process of spatial transformation was more complex than the gentrification processes typical of many US cities. This was not a displacement of one group by another, but a juxtaposition of multiple social groups, who occasionally coexisted peacefully but also fought amongst each other to define their own particular spatial capital. Cultural activities attracted a wide social spectrum, from local low-income residents to middle- and high-income groups, who had shunned the run-down Historic Centre for more than two decades, to artists who were encouraged by an increasing diversity of events and venues.

If we were to use the analogy of the Historic Centre as a theatre stage, we could identify the multiple spectators attending the many social and economic events. Spectators with season tickets would be those businesses and institutions who consistently had a presence at the Historic Centre and provided basic services in spite of the changing face of the area over the years, for example San Marcos University, the Ministry of Education, Cuneo shoemaking shop, or the Cordano

Bar. Seeking an exclusive historical reference in their consumption patterns and returning to the Historic Centre after a long absence, the upper income class would take the seat in the boxes. They would be attracted to the upscale cafés, to the opera at the plaza, or to a carefully rehabilitated *casonas*. Artists, professionals, merchants, and workers in the formal sector and middle class, would take the orchestra seats. They stepped in and out of the place according to specific circumstances and events. Many artists held jobs during the remodelling process and in public performances. Many lawyers, engineers, and government workers continued to serve local businesses and residents. Spectators with balcony seats would be former street vendors who were fortunate enough to relocate locally. They would be barely able to see the action on stage but at least they had seats in the theatre. Those seats would also be taken by the residents of the old *casonas* and *callejones* who lived in poor conditions and received very little direct benefit from recent changes at the Historic Centre. Outside the theatre, some spectators would choose first to read the review or watch the play on TV. They would want to know what was happening but from a distance.

This convergence of multiple social groups at the Historic Centre defines a major spatial transformation. The rehabilitation of public spaces and the development of cultural activities were more than a cosmetic fix. More importantly, this social convergence did not lead to the regurgitation of stale traditional cultural expressions but resulted in innovations that expressed the major social and political concerns of the time.

Politics of Place: The Collective Takeover of Public Space

Against this backdrop of economic change, physical improvement, and cultural activity, public demonstrations and political contest expanded dramatically at the Historic Centre. Many streets and plazas were used for gatherings by thousands of students, artists, workers, and mothers, among others, as their main stage for popular participation. Never before had the Historic Centre experienced this scale of political activity and its public spaces had never before projected such a strong sense of civic activism. It was evident that the appropriation of space by cultural workers and political actors had triggered a major expansion of spatial capital, rendering upon the Historic Centre a new urban vitality.

Multiple cultural-political events at the Historic Centre proposed a new relationship between civil society and the state. In the late 1990s and early 2000s, much of the protest against the Fujimori administration was expressed in subtle and powerful popular imagery, merging arts, culture, and politics, which was crucial in building the public participation that led to the collapse of that administration (Vich, 2002). These public events created new channels for citizen participation and political engagement in spite of the high level of corruption and government control over the media and the unprecedented levels of repression. Very simple acts, such as protest clothes laundering or writing, acquired national and international visibility when staged in proximity to powerful political and historic symbols in public spaces.

Artists at Work

Cultural workers and artists were at the forefront of this political process and engaged in the design of many politically charged marches, sit-ins and performances. In 1996, Víctor Delfín, a well-known creator of metal sculptures, initiated the collective artistic movement *Todas las Sangres, Todas las Artes*, or All Bloods, All Arts. This movement struggled against the dismissal of military abuses during the 1980s and 1990s. In 2000, it involved several other renowned artists in organizing two major successful events to protest against the fraudulent presidential elections that favoured Fujimori. The Fair for Democracy, staged at Campo de Marte, one of the largest open public spaces near the city centre, was characterized by the involvement of many non-governmental agencies in both organization and participation, and strengthened the opposition to the Fujimori regime. The Burial of the Corpse, an event that dramatized the death by corruption of the National Office of Electoral Vote, involved many performers, painters, and writers, and was staged in front of the Palace of Justice.

The *Muro de la Vergüenza* or Wall of Shame was another powerful political event organized by artist Roxana Cuba in protest against government corruption. It was originally installed at the Plaza San Martín and later relocated in front of the Palace of Justice. This wall was a 15-metre long cloth with a gallery of photographs of prominent political figures, from Vladimiro Montesinos, Fujimori's closest advisor, to Congresswoman Martha Chávez, Cardinal Cipriani, and Ambassador Francisco Tudela. The intention was to build a critical *dazibao*,[9] which served as a public space to express the rejection of the regime in power. This was a form of expression offered as an alternative to the co-opted media. These walls were replicated in many cities around the country between 2000 and 2002.

Plaza Mayor and the Small Towns in the Highlands

Plaza Mayor was not just a plaza for the Historic Centre or Metropolitan Lima, but it also became a plaza for the people of the small towns in the highlands. In September 1997, sixty municipal authorities from Huancavelica, one of the poorest provinces in the highlands, led by Pico Salas, the Mayor of the city of Huancavelica, marched to the presidential palace on Plaza Mayor to demand from the president a more equitable allocation of government resources for their neglected province. After an eight day journey, they arrived at Plaza Mayor riding their horses, dressed in their traditional and colourful attire, while local traditional music bands and dancers welcomed them. People from various communities joined them during their journey so by the time they arrived in Lima the procession numbered more than 1,500 people. This march had huge symbolic impact and delivered a powerful political message that brought national attention to the agenda of these neglected municipal governments. Thus, Plaza Mayor served as an ideal stage for media coverage at national and international levels, which in turn empowered the Huancavelica authorities and its people.

The Family House Patio extended to the Plaza

As the general disillusionment with the Fujimori regime intensified, political demonstrations not only grew in number and size but also began addressing general and core themes such as the essence of the relationship between state and society. *Lava la Bandera* or the 'Washing the Flag' ritual was an action that redefined the use and symbolic meaning of public space (figure 5.7). This ritual consisted of washing dozens of flags with 'Bolívar' soap in red washtubs[10] and then hanging them to dry on a line around the entire Plaza Mayor. The display of the long line with many flags around the plaza created an unexpectedly vibrant imagery, and involved a continuously growing number of people in front of the presidential palace every Friday between noon and 3 pm. The protesters demanded a major overhaul of government operations by the re-appropriation of a symbol that had been hijacked by the authoritarian ruler. The first few gatherings to 'wash the flag' were broken up by police, who called the ritual cheap symbolism. However, the ritual continued and was embraced by people from various social groups. It became very popular and was replicated in many plazas around the country with the addition of other washing rituals, such as military and religious uniforms. The day Fujimori resigned from the presidential office, the flag was being washed in twenty-seven cities around the country and several cities abroad.

According to Gustavo Buntinx (2001), writer, curator and one of the main leaders of this event, 'washing the flag' tapped into an extraordinary source of symbolic capital in the strategic reorganization of the democratic forces fighting the corruption of the Fujimori regime. The ritual combined, in bringing together washing and the flag, one of the most essential domestic functions with the collective re-appropriation of one of the most important national symbols. Public flag washing turned the plaza into an extension of the family house patio, and in so doing challenged the representational value of government power. 'Washing the flag' at the plaza allowed citizens to re-appropriate their public spaces, and

Figure 5.7. 'Wash the Flag' campaign at Lima's Mayor Square. (*Photo*: *El Peruano*, Víctor Palomino, 24 November 2000)

was a powerful symbolic act demanding a rethinking of the relationship between citizen and city, private and public, and arts and politics.

Popular Gathering Overwhelming Official Events

The largest public gathering to make use of the spatial capital of the Historic Centre was the *Marcha de los Cuatro Suyos*, or March of the Four *Suyos*, which took place during national Independence Day, 28 July 2000 (figure 5.8). This was a national political demonstration to protest against the Fujimori regime and its attempt to cling to power. The term *Suyos* refers to the major regions into which the country was divided during Inca times. About 40,000 people came from each of the four Suyos and converged on the centre. Another 250,000 people from the city of Lima itself gathered around Paseo de la República, one of the major boulevards

Figure 5.8. Demonstrators converging on the Cuatro Suyos March at San Martin Square. (*Photo*: *Diario El Peruano*, Susana Ayanoa, 28 July 2000)

in the Historic Centre, which was practically invaded by demonstrators who came from all corners of the city and the country, turning the previously fragmented opposition to the Fujimori regime into one consolidated front. The *Marcha de los Cuatro Suyos* demonstrated to national and international audiences that the democratic opposition could mobilize huge masses on the streets and plazas in addition to registering high figures at the polls. On 27 July, the eve of the national day, great political festivity in Paseo de la República announced the arrival of a new era. This party dwarfed the official events organized by the government at Plaza Mayor. While Plaza Mayor was guarded like a military fort with limited access to the general public, Paseo de la República was the stage for an open and spontaneous gathering with people flowing in and out, unimpeded by the restrictions. The government definition of a centralized and authoritarian space was overwhelmed by a public definition of a space imbued with a regional and national symbolic connotation. On this day, Paseo de la República turned into a space for a national gathering.

The powerful messages delivered by the people in the streets and plazas of the Historic Centre and other public spaces directly influenced the ensuing major political upheavals. President Fujimori resigned from the presidency, fled, and sought asylum in Japan.[11] Multiple legal charges were raised against him and arrest warrants were issued. Vladimiro Montesinos, Fujimori's closest advisor, was jailed, and for the first time in the nation's history a concerted campaign to bring former government and military officials to justice was under way.

The Changing Horizons of Spatial Capital

This study of Metropolitan Lima's Historic Centre illustrates a particular example of the production of spatial capital. The high profile and sophisticated cultural-political events, aligned with the qualities of the physical setting itself, together with the impacts of the events at many geographical levels clearly depict the dimensions of this spatial capital. The production of the capital was driven by the municipal rehabilitation programme, the social euphoria for the use of public space, the proliferation of cultural activities, and the staging of political demonstrations. The emergence of cultural-political events illustrates the power of spatial capital, instrumental in opening the floodgates to political debate and fostering popular participation in deposing a corrupt government. Collective participation was charged with strong historical symbols and cultural references and was staged in key public spaces. The central government lost control of the public discourse, which in turn redefined the meaning of the most prominent public spaces.

The political and cultural actors appropriated the historic and rehabilitated built environment to project powerful images to the city and the country through multiple social and media networks. The form, function and representation of the built environment provided the basic platform for a new set of urban activities. The intensity and diversity of the cultural-political activities defined the new role of the Historic Centre. The projection of those activities through local and international networks amplified the new urban vitality and its identity in the national and international context. In short, the alignment of actors and activities with the built environment and networks shaped the accumulation of spatial capital at the Historic Centre.

The new actors who had vested interests in the Historic Centre were, however, not a homogeneous group in any way. On the contrary, they had diverse visions ranging from upscale pedestrian streets with cafés and galleries, to a crowded plaza on which to demonstrate and demand better funding for small towns in the highlands. These visions were sometimes complementary, at others they clashed. However, they all gravitated towards the history and culture of the place. This place is impregnated with images of events and symbols that define the identities of Peruvian society. The *mudéjar* balcony brought from the Arab world by the Spaniards, Plaza Mayor born out of the gathering place established by Taulichusco, or the Paseo de la República's urban design influenced by Haussmann and implemented by President Piérola; they were all important physical elements that defined a perfect setting for cultural activities.

The spatial capital of the Historic Centre allowed the coalescence of major political powers, which ultimately triggered major changes in national government.

However, more recent events at the Historic Centre highlight the risk of losing this spatial capital, succumbing to the forces of urban chaos or becoming akin to a deteriorated museum artefact. This latent spatial capital must have raised the fears and concerns among local government authorities as the municipal administration of Castañeda, which followed Andrade's, prohibited political demonstrations in the core of the Historic Centre. This action could have the adverse effect of diminishing the role of the centre as the new space of social convergence, centre and symbol of a collective memory shared in a democratic process. It could very well amplify the fragmentation of a precariously united Peruvian society, where the unified centre might become a cultural anachronism. This restriction also highlights the tensions between the proponents of preserving the centre as a stage for entertainment and cultural activities and others advocating its role as a site for political expression. This is a struggle between two visions, one for a museum in which to observe history, and the other for a place in which to create history. Neither approach necessarily excludes the other, but these are the social tensions embedded in this locus of multiple and overlapping interests and voices.

Notes

1. Between 1950 and 1970, massive migration from the countryside resulted in a five-fold increase in Lima's population, from 645,200 to 3,302,500, and also major suburban growth through squatter settlements.
2. People of mixed European and indigenous ancestry.
3. Between 1987 and 1992, the gross national product declined more than 22 per cent, while in 1990, the annual rate of inflation was more than 7,600 per cent. The prevalence of poverty expanded dramatically, engulfing more than half of the population, and the unfortunate outbreak of a cholera epidemic caused thousands of deaths in the poorest segments of the population (INEI, 1997; Iguíñiz, 1996; Crabtree and Thomas, 1998).
4. The increase in GNP in 1994 reached 13 per cent, one of the highest rates of growth in many decades and one of the highest worldwide. Between 1990 and 1997, exports increased twofold and imports tripled (Banco Central de Reserva del Perú, 1998).
5. In 1990, the Historic Centre and its surroundings contained institutions that managed close to 50 per cent of all deposits and loans while San Isidro had about 30 per cent. In 1997, the share of the Historic Centre declined to 29 per cent and San Isidro's rose to 39 per cent (Chion, 2000).
6. Lima grew from 1.8 million in 1961 to 6.5 million in 1993 (Censo de Vivienda y Poblacion, Lima, Peru).
7. In 2003, Mega Plaza and Royal Plaza generated US$130 million in revenues (*The Economist*, 2004).
8. Architectural and design style developed in southern Spain with a strong Moorish influence.
9. Chinese wall journal, usually handwritten, showed in public places.
10. Bolívar, the soap brand, named after Simón Bolívar, a leader of the fight for independence in several South American countries. The red washtub is a reference to one of the national colours.
11. In November 2005, Fujimori travelled to Chile where he was arrested. In 2007 he was extradited to Peru where he is now incarcerated.

References

Amin, Ash and Thrift, Nigel (1994) Living the global and holding down the global, in Amin, A. and Thrift, N. (eds.) *Globalization, Institutions, and Regional Development in Europe*. Oxford: Oxford University Press.

Andrade, Alberto (1996) *Memoria Gestión 1996*. Lima: Municipalidad Metropolitana de Lima.

Banco Central de Reserva del Perú (1998) *Memoria Annual*. Lima.

Buntinx, Gustavo (2001) Lava la bandera: El Colectivo Sociedad Civil y el derrocamiento cultural de la dictadura en el Perú. Ponencia. Lima: Text Unpublished.

Castells, Manuel (1979) *The Urban Question: A Marxist Approach*. Cambridge, MA: MIT Press.

Castells, Manuel (1996) *The Rise of the Network Society*. Oxford: Blackwell.

Chávez, Eliana *et al.* (1998) *Perú: El Sector Informal Frente al Reto de la Modernización*. Lima: Oficina Internacional del Trabajo.

Chion, Miriam (2000) Global Links and Spatial Transformation in Metropolitan Regions, Lima in the Nineteen Nineties. Doctoral Dissertation, University of California, Berkeley.

Crabtree, John and Thomas, Jim (eds.) (1998) *Fujimori's Perú: The Political Economy*. London: Institute of Latin American Studies.

De Soto, Hernando (1986) *El Otro Sendero: La Revolución Informal*. Lima: Instituto Libertad y Democracia.

Günther Doering, Juan and Lohmann Villena, Guillermo (1992) *Lima*. Madrid: Editorial MAPFRE SA.

Hall, Peter and Castells, Manuel (1994) *Technopoles of the World: Making of 21st-century Industrial Complexes*. London: Routledge.

Iguíñiz, Javier (1996) *Empleo y Descentralización en el Perú del Siglo XXI*. Lima: Instituto Bartolomé de las Casas.

(INEI) Instituto Nacional de Estadística e Informática (1997) *La Actividad Económica en Lima Metropolitana*. Lima: INEI.

Instituto Libertad y Democracia, (1985) *El Otro Sendero: Datos y Análisis*. Lima.

Lefebvre, Henri (1991) *The Production of Space*. Oxford: Blackwell.

Ludeña Urquizo, Wiley (1996) A Propósito de los Patrones de Asentamiento en la Investigación Urbana de Lima. *Arquitextos*, October.

Ludeña Urquizo, Wiley (2002) Lima: poder, centro y centralidad. Del centro nativo al centro neoliberal. *EURE (Revista Latinoamericana de Estudios Urbanos Regionales)*, **28**(83).

Massey, Doreen B. (1994) A global sense of place, in *Space, Place, and Gender*. Minneapolis, MN: University of Minnesota Press.

Massey, Doreen B. (2005) *For Space*. Thousand Oaks, CA: Sage.

Ohmae, Kenichi (1996) *End of the Nation State, The Rise of Regional Economies*. New York, NY: Free Press.

Sassen, Saskia (1994) *Cities in a World Economy*. Thousand Oaks, CA: Pine Forge Press.

Smith, Michael (2000) *Transnational Urbanism, Locating Globalization*. Oxford: Blackwell.

Távara, José (1994) *Cooperando Para Competir: Redes de Producción en la Pequeña Industria Peruana*. Lima: DESCO.

The Economist (2004) Go north, Limeno, Mega Plaza and Royal Plaza. May 13.

Urry, John (1995) *Consuming Places*. London: Routledge.

Vich, Victor (2002) *Desobediencia simbólica. Performance y política al final de la dictadura fujimorista*. Buenos Aires: Consejo Latinoamericano de Ciencias Sociales CLACSO, Instituto de Desarrollo Económico Social.

Chapter 6

The Plaza de Bolívar of Bogotá: Uniqueness of Place, Multiplicity of Events

Alberto Saldarriaga Roa

Urban space, from the very moment of construction, acquires 'life' – the life of the citizenry that occupies it. Correspondingly, civic life appropriates urban space and incorporates it within its activities and its social imaginary. Thus conceived, urban space has a meaning in civic life that is strongly linked to the sense of what is public in society. This meaning is not static, it evolves over time. So put, the meaning of an historic urban space is an accumulation of past events with contributions from the present. The character of the physical space, as an urban void, also evolves; the buildings surrounding it may be replaced, thus what they connote evolves too.

In urban space, functional elements combine with symbolic ones to form structure and meaning, for instance, when ordering of the circulation of pedestrians and vehicles combines with symbolic elements related to the meaning of the public domain and the formation and reproduction of collective urban memories. A space can also be inclusive or exclusive, open to civic participation or reserved for special or exclusive groups. Alternatively, the strength of the notion of public in civic consciousness is reflected in the right to the use of urban space. Its weakness is evident when public space is surrendered to those who are able to exert power over it in one way or another.

This chapter is an analysis, from three different perspectives, of one historic urban space, the Plaza de Bolívar of Bogotá, and the transformation of its symbolism over time. The first is a theoretical foundation that refers to the sense of the public in urban space. The second is historic and deals with significant aspects of life in the plaza. Finally, I embark on a narrative analysis of the ordinary and extraordinary events that have taken place in the plaza, which illustrate different periods in the history of the city of Bogotá. The first section of the chapter proposes an interpretation of 'publicness' in the contemporary city, which takes us back to a tragic event that took place 20 years ago. There follows an historical account of some characteristics of what once was the main plaza of a small colonial city and today is the principal plaza of a city of some eight million inhabitants and the capital of a nation. I end this analysis by recounting some remarkable events of the twentieth century. In the final section, I illustrate the current state and uses of Plaza de Bolívar of Bogotá, particularly our thematic interest in the blurring of boundaries between everyday events and extraordinary political and artistic expressions of various sorts. It is important here to highlight that Colombia is a country that has been suffering from violent armed conflict for decades, a conflict whose effects permeate all aspects of urban life. Thus, the Plaza de Bolívar has witnessed tragedies, but also events whose aim is to transcend the ongoing drama in the search for hope for a peaceful existence.

The Urban Space, the Structure of the City and the Sense of the Public

The history of cities is, in a certain sense, the history of their urban spaces. Urban life – the aggregation of citizens' actions which take place in the public sphere and which find in public space the necessary stage – can nevertheless be either facilitated or restricted in urban spaces. An inextricable relationship exists between urban life and public space. The availability of spaces for meeting and interacting makes the city a living entity. The street as a space for movement and the plaza as a gathering space are two fundamental elements in the structure of the city. They assume different names, forms, sizes, and meanings in different historical moments and locations. On the streets, the flows of people and vehicles are channelled, while in the plaza citizens can gather in large numbers in order to celebrate, protest, participate, or dissent. As historians tell us, plazas were already present in the urban settlements of Mesopotamia and the Indus Valley (Frankfort, 1950).

The main plaza of a city is where both space and urban life are produced; the large open spaces of the past are its antecedents. One of these, the Athenian Agora, is rhetorically mentioned as the paradigmatic space of democracy, despite the fact that studies show the restrictions imposed on the participation of certain groups of citizens in Athens (Sennett, 1994, chapter 1). Alternatively, the ceremonial spaces of pre-Hispanic cultures were places for the mass congregation of people for the rituals, sometimes bloody ones, of their religious cults. The plaza has been and continues to be the appropriate place for all kinds of public activities. Urban planning, understood in its physical dimension as the ordering of urban space, does not necessarily claim an understanding of urban life as part of its conceptual formulations or its projects, this is habitually construed as the realm of the social sciences, removed from the professional expertise of the urban planner or architect. Yet, urban life inevitably flourishes in new and existing spaces, and it is therefore imperative to understand and address the sense of the public that enlivens those spaces.

In order for an urban space to become public, not only is its material existence required, but also its correlation with a notion that can be expressed as the *sense of the public*. This sense is not the same for all citizens as it has transformed over the years, and it is unlikely that it will remain constant in the future. So the public nature of urban space is not a problem that is based solely on its physical characteristics. For it to function as public space requires the expansion and strengthening of citizenship and of the sense of the public among all social classes inhabiting a city. The world of the public is defined in different ways. In legal terms, a public entity is something that belongs to the state, whereas private is something pertaining to individuals or particular entities. 'Public' is something that belongs 'to the people or that is related to them' and, therefore, is a common good. From a communicational perspective, public is something that is 'seen, understood and known by all', whereas private is 'something that pertains or is reserved for a single person or a limited and selected number of individuals'. Correspondingly, there are matters of public interest or public knowledge, while there are others that are exclusively of private, personal, or intimate interest. These conceptual distinctions between public and private are, nonetheless, difficult to ascertain in practice, and their boundaries have become ever more contested and blurred.

Furthermore, the public can be conceived of as a wide realm where identity and social networks are established through participation, a fostering of a sense of belonging, and outright appropriation. These transactions are related to material or non-material matters, including knowledge and information. In the Spanish language however, *public* is also a group of persons who attend an act or show, referring in this case to a group that performs predominantly passive contemplation of activities. The sense of the public for the purpose of our study of contemporary urban space assumes many of these meanings: it qualifies it as a common good; as a space where all citizens have a legitimate right to be; and furthermore, as a location where the spectacle of urban life takes place, and where citizens are simultaneously actors and spectators. The sense of the public and the development of citizenship have a direct and inevitable relationship, which is expressed in many ways in urban space. Citizens identify themselves through actions in which they participate in response to issues of collective concern. Participation has a connotation of action, and it is expressed in concrete events. However, in many aspects of contemporary political culture, participation means *attending*, or being spectators of what is being offered as a show. In the urban space of any contemporary city, there are events that involve an active citizenry and others where citizens are simply spectators. This raises a notable difference between the two basic senses of the public. In the first sense, we speak of the active participation in the common life of the community, and in the other, we speak of the passive contemplation of something that takes place in urban space, e.g., the spectacle of power or of art.[1]

Urban space is thus *par excellence* the space of and for citizenship. Special spaces, such as the main square of a city, assume an historical role in the formation of the citizens' consciousness. This role evolves over time. A question arises at this point: can an historical space, such as the Plaza de Bolívar, maintain a central position in the consciousness of today's citizenry?

Interlude: Two Special Moments in the Life of the Plaza

At midday on 7 November 1985 the Plaza de Bolívar in Bogotá was completely empty. Only the sound of rifles and machine guns broke the silence. Suddenly, a tank entered the plaza from one corner, climbed the stairs of the Palace of Justice, and without hesitation, entered the building destroying the main entrance. Hours before that, a missile had scarred the stone walls of the palace. The state was compromised, and something needed to be done. On the morning of the previous day, at 11.40 am, a group of 35 guerrillas, women and men members of the M-19 Movement, seized the Palace of Justice and held hostage those inside, including the magistrates of the Supreme Court of Justice of Colombia.

The still-incomplete Palace of Justice was situated in the northern section of the Plaza de Bolívar and had been inaugurated 15 years earlier. It housed the Courts of Justice of the Colombian nation. Its architectural design was very unique. The first floor was entirely enclosed with thick walls covered in limestone. The upper floors were open to the exterior with large narrow windows separated by stonework. The guerrilla soldiers entered through the parking lot and quickly took control of the building, which had only one access door situated directly in front of the plaza. The closed nature of the building helped the guerrillas' control of the building. Indeed, they were able to convert it into an almost impregnable fort. The tank's entrance into the building and the following events culminated in a tragic takeover. Twenty-eight hours after the initial actions, the palace was in flames, and by the next day only charred ruins remained. Many people died, their number is still unknown. Others simply disappeared.

On 6 November 2002, the Colombian artist Doris Salcedo carried out a unique artistic event to commemorate the tragedy of 1985. At 11.40 am four hundred wooden chairs were taken, one at a time, slowly, from one of the corners of the new Palace of Justice. The chairs were meant as a reminder of that tragic event, and the number approximated the number assumed dead that day during the guerrilla takeover. Initially, passers-by did not know what was happening. Then, they began to congregate in front of the Palace of Justice. They gathered in small numbers at first, but hours later they gathered in ever larger numbers. Rumours spread that the artistic event had something to do with what had happened in the past. The chairs were displayed together for a brief period of time, after which they were removed one by one. The event lasted 28 hours. The multitude of bystanders dispersed in the same fashion. Art had transformed the memory of the tragedy into an object of political consciousness.

The Plaza Mayor and the Colonial City

In the sixteenth century, the Spanish Crown imported an urban model to Latin America, which placed the plaza at the centre of the urban space and life in society. This model was used over the next three centuries as a foundation for the many cities that were to make up the continent-wide urban network of Latin America. The main plaza became the centre of urban life in cities and towns during the colonial era in Spanish America.

In the various urban plans for the first Spanish settlements in America, the main plazas were given a central location in a somewhat arbitrary manner. By 1520, norms to regulate urbanization were decreed. These decrees culminated in the *Ordenanzas de Poblaciones*, which were ordered by Felipe II of Spain in 1573, also known as the Laws of the Indies. In these laws, the plans for cities were designed in a much more structured manner and the plazas were located centrally. The most representative buildings of the church and the Spanish Crown, but also the residences of the most powerful inhabitants were located around the plazas. The roads connecting the colonized territories terminated in these plazas.[2] Thus, the plazas became the centre for both ordinary and extraordinary events.

Since then, the main plaza has been a symbol of historic Hispanic American urbanism. The expansion of metropolitan centres has, however, diluted that status. In villages, towns and even in small cities, plazas keep aspects of their original character. In large urban centres, the situation has changed. The main plazas, which in the past were the centres of small urban worlds, today are to some extent lost in the extensive urban fabric of metropolitan areas. In this light, new questions emerge: How do these plazas function, both in terms of their use and meaning, in the context of large Latin American cities? Have they acquired new functions and meanings? As we shall see, the case of the Plaza de Bolívar in Bogotá can provide some answers to these questions.

The Plaza de Bolívar and its History

There is no official record of the main plaza of the city of Santafe, which was Bogotá's original name.[3] The officially recognized date of the city's founding is 6 August 1538, but it was in April 1539 that the plans for the city were defined and a Council was named to govern it. Both the perimeter of the plaza and the allocation of its surrounding lands date from that year (Martínez Jiménez, 1988). The original plan for the distribution of land was lost in the fire of 1900, which destroyed part of the city's historical archives. In 1550, the Spanish Crown designated Santafe de Bogotá the capital of the New Kingdom of Granada, centre of the regional Archbishopric, site of the Catholic Church, and location of the Royal Audience. This spurred the construction of the first buildings surrounding the plaza: the Cathedral and the Seat of the Royal Audience. The construction of the first cathedral began in 1553. The church, however, collapsed twice over the years; first in 1566 (it was rebuilt in 1585), and again in 1785 and was finally replaced with the cathedral that stands today, which was designed by the Spanish priest Domingo de Petrés. The House of the Royal Audience remained until 1846 when it was demolished in order to make room for the new Palace of Government, currently known as the National Capitol.

Everyday life in Santafe de Bogotá during the colonial period evolved around events relating to public administration, religion, and education. Religious and political celebrations were, for almost three centuries, the events that made for variety in the otherwise routine life of a small city located in the highlands, at

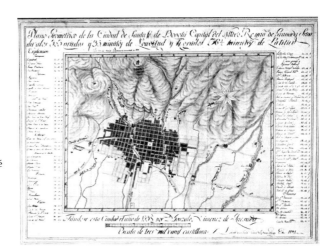

Figure 6.1. Plan of Santafé in 1791, by Engineer Domingo Esquiaqui. Reproduced in Madrid in 1816. The original is in the Archivo Histórico Militar in Madrid, Spain.

an altitude of 2600 metres. The Catholic Church's complex religious calendar set a routine of prayers, masses, and festivals. Certain secular events, such as the reception of Spanish governors, the birthdays of the King, the Queen, and the Princes of Spain, and popular celebrations such as bullfights, cockfights and carnivals were enmeshed in this web of religious events.

During the colonial period in Santafe, there was a great proliferation of celebrations. The reception of a viceroy or an archbishop, or other events of certain scale were often measured according to the days dedicated to celebrate them. As a matter of fact, those celebrations could last for several days. For instance, those events could last up to 15 days, during which the feeble activity of the city should come to a halt. That very relaxing notion of time was a very intricate part of the spirit of the times. Not until the Borbonic period, towards the later portion of the colonial period, was there an attempt to regulate the use of time through utilitarian functions. Since 1747, small efforts were perceived to reduce the number of celebrations and their duration. In response to the courts' lack of efficiency, a royal decree declared the reduction of the 'Celebrations of the Court' during which people did not work, keeping only the 'Celebrations of [religious] Precept'. (Vargas Lesmes, 1990, pp. 303–304)

The numbers tallied in Julian Vargas Lesmes's 1990 work identify nearly 50 days of religious celebrations per year in the city of Santafe de Bogotá. Apart from those celebrations, the author added 'half celebrations' and the extraordinary 'table celebrations' of greater political significance. Civil and religious authorities presided over these celebrations from the atrium of the Cathedral, on the eastern side of the plaza (Vargas Lesmes, 1990, pp. 304–305). Thus a sixth of the year was dedicated to celebrations that had the space of the Plaza as their main stage. The celebration of Corpus Christi or of the Eucharist was, according to various reports, the most spectacular in Santafe de Bogotá and was celebrated throughout the nineteenth century in Bogotá – the name officially given to the city after the triumph of the patriots in 1819. A French traveller, Gaspard-Théodore Mollien, described that celebration in the following way:

The feast of Corpus Christi is the most spectacular celebration in Bogotá. The event's eve is announced with fireworks. Four altars are erected and richly adorned, one on each corner of the main plaza through which the procession will pass and, because of a unique mixture of the sacred and the profane, crosses are placed throughout the area. Games, puppetries, and exotic, caged animals were also placed around the area. The games cease just as soon as the bell sounds to announce the arrival of the procession. Everyone at this point unmasks himself and kneels on the street. (Mollien, 1992, p. 224)

Bullfights, that were originally held during the pagan celebrations of earlier times, have been held in the principal plaza since the sixteenth century. For the occasion, a bullring and tiers of seats around the perimeter of the plaza were constructed. Certain bullfights lasted several days, to the misfortune of vendors who had their stores in the plaza (Vargas Lesmes, 1990, p. 312).

Independence and its Effects on the Plaza de Bolívar

An extraordinary event, the Proclamation of Independence, took place in one corner of the main plaza of Santafe on 20 July 1810. The incident has been described in various ways in the history books of Colombia and has been meticulously analysed in historiographic studies. According to the most popular version, a citizen of Spanish descent named Luis de Rubio engaged in a dispute with a Spanish businessman named Francisco Llorente because the businessman had refused to lend him a porcelain vase for a celebration. The incident would usually have gone unnoticed, but a citizen that happened to witness the argument and kindly greeted the Spaniard was nevertheless insulted by him. This insult angered other pedestrians who were passing and they rapidly gathered to protest against the presence of Spaniards. In a couple of hours, a few well known leaders, who had with cessionary views, took advantage of the popular upheaval to establish a new government in the city. In the same short time that it took to form the popular movement, a declaration of Independence was drafted, and publicly proclaimed at the Plaza Mayor at six in the evening the same day. Historical research shows that this event acted as a catalyst that speeded the ramifications of serious political tensions that had been evolving for many years (Alba, 1989). This was the starting point of a long and bloody war of Independence that lasted 9 years. During this period the plaza was the scene of public executions and the hanging of patriots who were punished by the Spanish Crown. For several years, Santafe experienced the so-called 'regime of terror' established by the Spanish General Pablo Morillo, known as The Pacifier.

The Republic of Colombia was finally established after 7 August 1819, the day on which the Battle of Boyacá put an end to the war of Independence. A few months later, Simon Bolívar, the Liberator, arrived in Santafe as the President of the new Republic. The Plaza Mayor was the place of his public arrival. From that year, the city was called Bogotá and became the capital of the new Republic. In the years after Independence, Bolívar's name was ascribed to many urban spaces, streets, plazas, avenues and parks, in the five Latin American countries freed from Spanish rule by his leadership of opposing forces. Thus, the Plaza Mayor was renamed as the Plaza de Bolívar.

Many political, religious and cultural events took place in the Plaza during the nineteenth century. A significant number of them were the consequence of the civil wars and violent regional conflicts of that century and included battles, executions, the reception of victorious troops, and religious and memorial services. Other more peaceful events also took place, for example the inauguration of the statue of Simon Bolívar that was shipped from Europe in 1846 and placed in the centre of the Plaza, in the position formerly occupied by the colonial public fountain. A weekly market was held in the plaza until 1864 when, for sanitary reasons, it was moved to a new location. Years later, in 1881, a park was created around the statue of Bolívar, which was enclosed with an iron fence. This change created new customs among the citizenry. The old gathering place was divided, its centre was isolated, and activities were displaced to the edges of the plaza. One of the peripheral locations, the cathedral atrium, took on special importance:

> Altozano is a Bogotanian word that simply designates the cathedral atrium, which occupies one side of the Plaza de Bolívar. The atrium rests on five or six steps and is 10 to 15 metres wide. There, during the morning, after taking the sun whose ardour is eased by the fresh ambience of the height; and during the afternoon, from five to seven, after dinner (the Bogotanian eats at four), all remarkable people in politics, literature, or position, meet daily.
>
> An exchange, a literary circle, an *areopago*, a *coterie*, a singles bar, a theatre *coulisse*, a forum, all of Bogotá's activities in a hundred square metres: that is the *altozano*. If the silent walls of the church could speak, they would certainly tell the history of Colombia, from the etiquette struggles of delegates and bishops of the Colony, from the chronicles of *El Carnero Bogotano* until the latest conspiracies and uprisings! More than once, blood has stained the stones of its pavement, more often than once, it has been the theatre of savage struggles. The citizen of Bogotá is attached to the *altozano* because of its intellectual atmosphere, because there thousands gather ready to hear and taste a spiritual occurrence and watch the four winds pass by (Cané, 1992, pp. 153–155).

According to Miguel Cané (1992), the *altozano* was the social club of Bogotá. That tradition had been established from the beginning of the nineteenth century, as has been confirmed by other travellers' stories. The atrium was reserved for men. Women who visited it in the afternoon were frowned upon. Broadly speaking, the presence of a lone woman in a public space was unthinkable. To go shopping, women had to be escorted by their maids. On Sundays, when families strolled together, women were afforded greater liberty to walk around the city.

Figure 6.2. Military parade in the Plaza de Bolívar circa 1900. (*Source*: Sociedad de Mejoras y Ornato de Bogotá; unknown photographer)

By the beginning of the twentieth century, Bogotá had already grown beyond its colonial borders. The Plaza de Bolívar kept its centrality with regard to religious and political activities for a number of decades after which its role changed for the life of the citizenry. Its physical space had been transformed considerably by 1900. Only the eastern part retained any semblance of its original appearance. The Capitol, which was erected along the southern edge, was still under construction. The Galleries, a commercial building that had occupied the western perimeter since 1846, had burned down in 1900 and was soon replaced by a new French-styled building. Today this building houses the Alcaldia Mayor de Bogotá (the City Hall). Other events also occurred in the Plaza. The country's presidents used to celebrate their inauguration on the steps of the Capitol. Luxuriously dressed bishops presided over traditional religious ceremonies. The citizens visited the plaza and occasionally gathered to voice their discontent over political matters. Mule-drawn trolley cars circulated around two of the plaza's sides from the 1880s.

The plaza, turned into a park in 1881, was remodelled again in 1926. The level of the central space was lowered and four large fountains were installed in each corner. By doing that, its role as a gathering place was further constrained and its

Figure 6.3. The Plaza de Bolívar as park. (*Source*: Sociedad de Mejoras y Ornato de Bogotá. (*Photo*: Henri Duperly, 1895)

Figure 6.4. The fountains of the Plaza de Bolívar built in 1926. (Coloured postcard, 1945; private collection)

contemplative character was strengthened. Urban life surrounded the plaza, but did not live within it, except during political events when multitudes gathered to support or reject a politician or a political party. All this was brought to a sudden halt on 9 April 1948.

A Turn in History

Jorge Eliécer Gaitán was born in Bogotá in 1903. He began his political career in 1930 after studying law in Bogotá and Rome. He quickly became the most popular leader in Colombia. As a lawyer of the dispossessed and an opponent of the dominant oligarchy, Gaitán built up an alternative political project that troubled the ruling class. In contrast, his revolutionary ideals and his vigorous discourse were enthusiastically received by the popular masses. They participated in demonstrations that were unprecedented in Bogotá's history. In one of these massive gatherings, 'The March of Silence', nearly 100,000 citizens illuminated the plaza with torches on the night of 7 February 1948 in a moving protest demonstration. Silence replaced the usual noise that had characterized these sorts of events in the not so distant past.

Gaitán was assassinated on 9 April 1948 at 1.00 pm. The riots in the wake of his death started almost immediately and lasted for several days; hundreds of citizens died as a consequence. Many buildings in the historic centre were burned down. Right wing politicians capitalized on the political uncertainty to establish a rightist regime, all of which made the situation much worse. The Plaza de Bolívar became the stage for enraged battles between government forces and angry masses. There are memorably tragic images of trolley cars in flames in front of the National Capitol and in front of the Cathedral. The city changed from this point on, as rural violence extended all through the nation causing the migration of hundreds of thousands of peasants to the cities, many of whom relocated on the periphery of the capital and other urban centres. New forms of urban life started to take shape.

The Ordinary and the Extraordinary in the Second Half of the Twentieth Century

The rapid and uncontrolled growth of Bogotá changed the significance and meaning of everyday life in the city, which became mechanized and acquired at least in part the qualities of modern city life. Some traditions of the past were preserved, along with new forms of use of public space. The Plaza de Bolívar, besides being the main plaza of Bogotá, was the principal space for the administrative and financial functions in the city and also retained its characteristic as a local plaza surrounded by neighbourhoods inhabited by the various segments of society. The routine life of the Plaza became a hybrid of its symbolic central role in religion and local and national politics. Its symbolic character combined with its use as a meeting point for public and bank employees, a must-see place for citizens and visitors of Bogotá, and a space for recreation for local families. Along with that, the plaza also has a role as an artistic and cultural space.

A special ritual that has taken place on the plaza since 1950 is the student parade from the schools of Bogotá, held yearly on 20 July as part of the celebrations of national Independence. However, the design of the plaza with its steps, fountains, and fences hindered any sort of organized formation. Furthermore, the edges of the plaza had been converted into parking spaces. For these and other reasons, a new remodelling plan was initiated, designed by the architect Fernando Martínez Sanabria. The new plaza was inaugurated with a student parade on 20 July 1960, 150 years after the declaration of Independence. The new design cleared the plaza of all obstacles, but retained the statue of Simón Bolívar on a new pedestal. The plaza regained its civic character, which had been lost in prior modifications. A new building for the Palace of Justice was constructed on the northern perimeter, and it was this building that was destroyed by missiles, tanks, and fire in 1985.

In 1991, the anthropologist María Clara Llano undertook some interesting research into everyday life on the Plaza de Bolívar. She captured fragments of what normally takes place there by direct observations and interviews. She found an 'official life', which includes everything connected with the political functions of institutions, such as the City Hall, Congress, the Court of Justice; a 'tourist life' of national and international visitors who recognize the plaza as an important place for the memory of the city and the country; and a 'local life', which pertains to the life of the street peddlers, shoe shiners, and the retired elderly who gather to recall old memories. Today, as it was in the past, storytellers and poets make use of the plaza for their performances during holidays. Soldiers and police officials continually guard the public buildings. It is in this everyday life that protest and artistic expressions, which have adopted the plaza as their preferred site, are sporadically entangled.

The Plaza de Bolívar has lost many of the functions that it performed during history. Yet, today one of its main functions is the representation of that history. School children visit the plaza daily to learn about the events of the independence; passers-by revive national history and some of the events that are part of its 'dark memory'; national and international tourists come to the plaza to get to know the heart of the Republic. The plaza is thus used as a monument, as a space from where the memory of the struggles of the nation's heroes, liberals, students, guerrillas, soldiers, and administrators are remembered. (Llano, 1991)

In August 2005, the Plaza de Bolívar of Bogotá, formerly Plaza Mayor of Santafe, celebrated its 467th anniversary. A space of such age has many stories to tell and has witnessed ordinary and extraordinary events, some pleasant, others turbulent and bloody. Each day that goes by something happens within the plaza, from the routine affairs of the ordinary lives of those who wander around it, to, occasionally, public protests denouncing crimes in violation of human rights in a country in conflict. An important centre within the extensive urban fabric that surrounds it and a venue for artistic and cultural festivities, the plaza still maintains its symbolic aura, and is a place of pilgrimage welcoming citizens and visitors alike.

The Space of the Plaza Today

An urban plaza is an empty space surrounded by buildings. The void acquires the character given it by the surrounding architecture and urban design. The Plaza de Bolívar is surrounded by buildings that are representative of different historical periods. The Cathedral, located on the eastern side, is a neo-classical building constructed at the beginning of the nineteenth century. To its side are the two oldest buildings on the plaza, the House of the Ecclesiastic Council and the Chapel of the Sagrario, which were constructed in the early seventeenth century. On the south side is the Capitol, an impressive neo-classical building designed by Thomas Reed in 1847. On the west side lies the Liévano building, the contemporary City Hall of Bogotá, a French neo-classical building designed by Gastón Lelarge at the beginning of the twentieth century. On the north side, there is the new Palace of Justice that replaced the one destroyed in 1985, which is the most recent building and also the one that does not quite fit within the plaza's general architectural style. The Plaza de Bolívar is a space where the predominant building materials are brick and stone. Its main buildings have façades carefully carved in limestone while some are faced in brick. The floor of the plaza is of the same materials. In spite of the changes throughout the centuries, the plaza acquired a coherent, orderly, and sober image. There are no traces of vegetation. Pigeons occupy the plaza during the day along with citizens; during the night, they sleep within the mouldings and crevices of the buildings which line it.

War and Peace: From Public Space to Artistic Space

It was first baroque art and then the neoclassical art of the nineteenth century that adorned the plaza with pronounced ornamentation, sculptures, and commemorative monuments. In the twentieth century, it was the time for modern art to make its mark in public space with its abstract geometries. Today, there are some artistic forms that are intimately identified with the space and life of the city. Doris Salcedo's work, mentioned at the beginning of this chapter, is a good example of this type of work and is particularly interesting for its political content.

It is precisely that combination of the political and the artistic that draws attention to the new role of the Plaza de Bolívar in the life of Bogotá.

In 1960, an experimental theatre company, the Theatre School of Cali, decided to perform Sophocles's tragedy 'Oedipus Rex' in the Plaza de Bolívar. The group of actors, who were directed by Enrique Buenaventura, selected the austere neoclassical buildings as the backdrop and the steps of the National Capitol as the perfect stage for the performance. In spite of the limited technical resources of the time, the voices of the performers were heard throughout the plaza along with the dissonant background music composed by Roberto Pineda Duque. That day, without much ceremony, the plaza was transformed from a public space almost exclusively reserved for religious and political celebrations to a space that was also for artistic performances.

In 1988, the Ibero-American Festival of Theatre was organized in Bogotá for the first time. During two weeks of April, this annual event brings together theatre groups from various countries of the world and includes dramatic performances in public spaces. The Plaza de Bolívar is now revered as the principal urban stage for theatrical performances. The first major performance, in 1996, was by the Catalan company Els Comediants. It was an event that made use of the entire space of the plaza where the actors symbolically represented the tragic events of 1985, and the performance included fireworks and even risky action scenes. Although in that performance the plaza ephemerally took on a special function, that extraordinary event in 1985 was imprinted on collective memory in a lasting fashion.

The armed conflict that Colombia has endured for several decades has made the Plaza de Bolívar the proper place for periodic collective manifestations of protest and mourning. For example, on 8 March each year, which is the International Day of Women, a great number of women come together to call for peace. And for 6 years now, the 'Exodus Expedition' commemorates the displacement of rural inhabitants and their migration to urban areas as a consequence of violence in Colombia. The definition of that 'Expedition', given by the organizers, is the following:

It is a cultural gathering of diverse organizations of people who have been displaced along with artists, women, and delegates of NGOs who work with this population and some institutions who have joined this initiative, which is six years old now. The Exodus is also a travel through different disciplines and languages in order to reinterpret this drama and give it the cultural dimension it deserves. The expedition will examine collectively the forced displacement in Colombia from the perspective of women, artists, actors, political reflections, personal testimonies, oral histories, and the arts. (Expedición Exodo, 2005)

As part of the Expedition's activities, on 21 November 2004 two performances were held at the Plaza de Bolívar. The first was entitled *Women and Suitcases* and was coordinated by theatre director Patricia Ariza. It was performed by 500 women dressed in black and white. The other performance, entitled *The Travelling Shoes*, was coordinated by Francisco Bustamante of the NGO Minga. It was performed by children who had been displaced accompanied by children of Bogotá. These events attracted entire families who had left their places of origin and migrated to Bogotá. Women and children paraded through the space of the plaza with empty suitcases. The sorrow of the displaced invaded for hours the space of the plaza, tainting with pain and sorrow the participants and spectators, but also opening doors for questioning the drama of violence and displacement in the country and the necessary reforms to overcome this tragedy.

Art and Politics in Urban Space: A New Sense of the Public?

This chapter has discussed some of the transformations in the citizens' understanding of public space that have happened over the four centuries of the Plaza de Bolívar's existence. What was deemed as public during the colonial period, greatly differs from the current sense of public. The dominant presence of the Spanish Crown and the Catholic Church shaped colonial life for centuries. Independence brought a sense of citizenship and democracy that was nuanced by the effects of class divisions and the corresponding segregation. Today, the plaza is a democratic space in which the problems of a society in conflict are presented and enacted. Public space is not just the space that belongs to all citizens, it is also the place where terrorist attacks and even open battles can happen. In a regime of fear, public space both attracts and frightens.

If the true sense of the public is active participation, then citizens should be involved in the decisions that affect them. For instance, they should be able to defend that which belongs to everybody, such as the urban space, and to be conscious of the public interest as a realm of communication and action. However, the realities of Colombia today show how the sense of the public

Figure 6.5. Citizens drinking hot chocolate in the Plaza de Bolívar, invited by the metropolitan mayor Antanas Mockus, 1995. (*Photo*: Alberto Saldarriaga Roa)

weakens and increasingly favours private interests. Indeed, in Colombia, reality demonstrates the ways in which networks of power strategize in order to pursue their objectives. The former is illustrated by the deliberate provocation of the *fear of the public*, used as a tool to support the use of repression, particularly in societies where the sense of citizenship is new and fragile and where the social conditions accentuate inequalities. This fear is spread through the exacerbation of trepidation towards the city and particularly towards urban space where 'everything is possible'. Conversely, private spaces are portrayed as venues of protection and security. While the space in the streets is presented as the site of unpredictable risks, shopping malls are perceived as ideal sites for consumption and recreation. In this sense, commercial centres represent progress as opposed to traditional streets, which are seen as backward.

All this is projected onto the construction of urban imaginaries, understood as the mental representations that citizens make of their cities. Bogotá is a stratified and segregated city. Each social group has a distinct image of the plaza. It is possible that a small portion of the seven million inhabitants of Bogotá has never visited the plaza, yet people have it as a point of reference in its imaginary. They even have a visual image of it, which they have obtained from historical accounts or mass media reports of the activities that took place and take place there. For others, the plaza is still a place where their needs are met: commercial goods bought, recreation sought, business transactions concluded, and even romantic rendezvous pursued. Public protests find in the plaza a vital space for their expression. The presence of the institutions of the state in the vicinity propitiates civic and governmental interactions.

It is important for the people of Bogotá to know that the Plaza de Bolívar is a space for all citizens, where each has the right to manifest his/her appreciation or dissent and views on political and economic conditions, along with the right to gather to appreciate cultural expressions which otherwise may be inaccessible. The image of the plaza as a space for protest is today tied to its function as a venue for artistic expression. The will of citizens to express dissent and their desire to attend cultural events can be intertwined in a single space.

The simultaneity of events is part of the new life of Plaza de Bolívar, and in some cases it attains extraordinary dimensions. While hundreds of women protest the disappearance and kidnappings of their family members, one can just as easily see pigeons flying around the plaza, a street vendor seeking to support his or her family, and police patrolling the Capitol to protect its safety. The plaza, public space *par excellence*, is witness to the authoritarian power that constraints the free flow of pedestrians. Art is at times employed to express citizens' defiance and in this way ratifies the value of a space where the life of the nation and the life of the city can be expressed.

Figure 6.6. Youth parade in the Plaza de Bolívar, 2004. (*Photo*: Alberto Saldarriaga Roa)

All of this gives importance and character to the Plaza de Bolívar. It maintains its original character as a symbol of the public. It is a space for recreation and religious rituals. The austerity of the buildings and the surface of the plaza induce a feeling of solemnity, which art is occasionally able to dispel. The plaza is a key centre for the everyday life of many, while the presence of guards gives it a militaristic image. Theatre converts it into a stage, transforms it into an acoustic venue, and recreates it as a space of spectacle of light and colour.

In a country where armed conflict casts its heavy shadow on daily life, art and protest are treasured gems that transform the violence in citizens' consciousness. They are insurgent demonstrations that contrast with most official propaganda and mainstream media. Their immediate physical and emotional impact is meaningful, while their deep effect in the citizens' consciousness is slower but more transcendental.

Figure 6.7. Artistic exhibit in the Plaza de Bolívar, 1999. (*Photo*: Alberto Saldarriaga Roa)

Epilogue

The citizenry that arrives at the Plaza de Bolívar expects to be immersed in a special experience. Its architecture is sober and invites solemnity. It is a space for everyday life, but also for extraordinary events. This has been both its historical and its current role. When hundreds of women gather to call for peace, they personify the country's tragedy and hope. When actors and artists occupy the plaza, they symbolize not only this tragedy and this hope, but also the very life of the city and the world. Through the enmeshing of ordinary and extraordinary events, past and present come together every day. This can be the real sense of a public space in the life of a contemporary city.

Notes

1. Richard Sennett has examined extensively the theme of what is public in urban space and its transformations in European cities since the eighteenth century. See Sennett (1975).
2. The most complete study published in Colombia is Salcedo (1996).
3. The original name provided by the founder Gonzalo Jiménez de Quesada was Santafe to which the additional name 'de Bogotá' was added. In certain texts the names are mistakenly separated: Santa Fe. See Martínez Jiménez (1988).

References

Alba, Gonzalo Hernández de (1989) La Nueva Granada en 1809 y 1810, in *Historia de Colombia*. Tomo 7. Bogotá: Salvat Editores, pp. 787–809.

Cané, Miguel (1992) *Notas de viaje sobre Venezuela y Colombia, 1881–1882*. Bogotá: Biblioteca V Centenario Colcultura, pp.153–155.

Expedición Exodo. http://www.expedicionexodo.com, accessed in 2005.

Frankfort, H. (1950) Town planning in ancient Mesopotamia. *Town Planning Review*, **31**(1), pp. 99–115.

Llano, María Clara (1991) Plaza de Bolívar. La manzana de la discorida. *Revista Proa*, No. 406, pp. 54–57.

Martínez Jiménez, Carlos (1988) *Santafé, capital del Nuevo Reino de Granada*. Bogotá DC: Fondo de Promoción de la Cultura, Banco Popular.

Mollien, Gaspard-Théodore (1992) *Viaje por la república de Colombia en 1823*. Bogotá: Biblioteca V Centenario Colcultura, p. 224.

Salcedo, Jaime (1996) *Urbanismo Hispano-americano. Siglos XVI, XVII y XVIII. El modelo urbano aplicado a la América española*. Bogotá: Centro Editorial Universidad Javeriana.

Sennett, Richard (1975) *The Fall of Public Man*. New York: Norton.

Sennett, Richard (1994) *Carne y Piedra. El cuerpo y la ciudad en la civilización occidental*. Barcelona: Ediciones Península.

Vargas Lesmes, Julián (1990) *La sociedad en Santafé colonial*. Bogotá DC: CINEP.

Chapter 7

Space, Revolution and Resistance: Ordinary Places and Extraordinary Events in Caracas

Clara Irazábal and John Foley

Introduction

As discussed in the introduction, many scholars have argued that public space is a prerequisite for the expression, representation, preservation and enhancement of democracy (Boudreau, 2000; Caldeira, 2000; Holston and Appadurai, 1999; Low, 2000; Sassen, 1996; Low and Smith, 2006). This has not been more true than in the capital cities of Latin America in recent decades, where political demonstrations have played a critical role in the demise of totalitarian regimes and the re-establishment of democracy. Caracas, capital city of Venezuela, is a prime example in that key urban spaces have been sites for popular demonstrations since Hugo Chávez became President in 1998. Public buildings such as the seats of political and military power, and private ones such as headquarters of media corporations functioned as architectural icons and background to these events.

While much attention has been paid to the role of social, political, and economic factors in the development of these events (Ellner and Hellinger, 2004; Márquez and Piñango, 2003; Méndez, 2004; McCaughan, 2004; Santamaría, 2004),[1] we know less about the influence of the physical and symbolic dimensions of public space.[2] We argue that the urban location and context of particular buildings and spaces, their accessibility to different groups from various areas of the city, and their relative symbolic value have been critical in the unfolding of political events, especially during the *coup d'état* and Chávez's subsequent reinstatement.

Caracas Scapes: Geographical and Socio-Political Fissures

Covering 912,050 square kilometres (slightly more than twice the size of California), Venezuela has a relatively small population of 25,375,281 people (July 2005 estimates). With the world's sixth largest oil reserves, the national economy is heavily dependent on petrochemicals.[3] However the proceeds from these resources have not trickled down, leaving 49.4 per cent of the population poor, 21.7 per cent indigent, and 71.1 per cent living below the poverty line (UN/CEPAL, 2001, p. 57), with all estimates showing a dramatic increase in poverty between 1975 and 1995. According to an income-based poverty measure (Riutort, 1999), 33 per cent lived in poverty in 1975 and 70 per cent in 1995. If poverty more than doubled in that period, the number of households in extreme poverty increased three-fold, from about 15 to 45 per cent. Though other non-income-based measures show slightly lower figures, Venezuela registered the largest increase in poverty and has the largest proportion of the population living in poverty in Latin America (Wilpert, 2003b).

It is no surprise that this notable inequality animates contemporary political struggle. In the country's main cities, such as Caracas, poverty is reflected in squatter settlements (popularly called barrios), accommodating over 40 per cent of the population.[4] Approximately 1,000 metres above sea level, Caracas's 5.1 million people (2004),[5] inhabit a dense narrow valley bordered by mountains at a density of 1,011.59 people per km^2 in a metropolitan area of 2,050 km^2. Many of the poor barrios are on steep hills, particularly in Petare in the east (also home to the middle and upper class) and Catia and La Vega in the west (mainly a working- and lower-middle class area). Barrios are sprung along river courses, out of sight until the occasional floods wreak havoc.

Right in the middle of the city are the curvaceous 'garden-suburbs' of the Caracas Country Club. In the east, high-density blocks of upper-middle-class apartments are spaced on regular streets or climb precariously up the hillsides. To the west, large public housing projects stand out, including the massive Le Corbusier-inspired '23rd of January' estate.[6] In various parts of the city, lower-middle-class apartment buildings crowd the narrow streets of the traditional settlements. On the hills, or snuggled next to exclusive villas of the wealthy in the east, are the intricate, maze-like streets, paths, and stairways of the barrios. The main bustling commercial and industrial core linearly traces the spine of the valley. The hodgepodge of uses and styles of architecture is visually united by the Avila rising another 1,000 metres to the north, as well as by the tropical vegetation sprouting out of any spare space.

Administratively and politically, Caracas is a fragmented city. The capital strides two large, politically independent entities, Miranda State and Capital District.

The Capital District, seat of national government, has only one municipality –
Libertador, whereas four municipalities – Chacao, Sucre, Baruta, and El Hatillo are
in Miranda state.[7] Additionally, a metropolitan level of government was created
in 2000 to coordinate the activities of the five municipalities. Approximately 65
per cent of the city's population is in Libertador and 23 per cent in Sucre, leaving
only 12 per cent in other three. Most *barrios* are located in Libertador (65 of the
barrio population) and Sucre, the two largest municipalities. Until 2004, these
two municipalities had mayors loyal to the national government, whereas Baruta,
Chacao and El Hatillo had mayors opposing it. Similarly, the Metropolitan Mayor
until 2004, Alfredo Peña, was an active opponent of the national government,
although he was elected with its support. In the 2004 elections, however, support
for Chávez's administration increased with the election of Juan Barreto as
Metropolitan Mayor.

These rifts were exacerbated by the neo-liberal economic policies implemented
by Venezuela's government throughout the 1980s. Together with political
corruption and mismanagement, they greatly reduced the standard of living.
The marked reduction in government spending was especially detrimental to
the poor, as per capita spending was reduced by 40 per cent between 1980 and
1993. Education spending was slashed by more than 40 per cent, housing and
urban development projects by 70 per cent, and health services by 37 per cent.
As a result, between 1984 and 1995, the poor doubled to 66 per cent of the total
population with those in extreme poverty rising from 11 per cent to 36 per cent
(Roberts, 2004). In 1982, 60 per cent of the population could count themselves as
part of Venezuela's privileged middle-class, but by 1990, only 34 per cent could
(Márquez and Piñango, 2003). The economy continued to deteriorate throughout
the 1993–2003 period (García-Guadilla *et al.*, 2004, p. 10).[8] The following sequences
of maps present a socio-economic and socio-spatial analysis of the metropolitan
area with data from 1998.[9] Predictably, these factors are fairly correlated with the
population's political preferences (shown in figure 7.5). This mapping helps us
understand the uses and meanings of public spaces in the dynamic reformulations
of democracy and citizenship in the city.

Figure 7.1 shows the five municipalities that compose the metropolis.
Representing population, the map shows that in effect, the two most populous
municipalities are Libertador and Sucre, governed by mayors who support
Chávez. Complementary, figure 7.2 shows the density in Caracas. Again, the
densest areas are in the east and west of the metropolis, coinciding with the
greatest concentrations of popular, mostly self-built *barrios* and low to middle-low
income areas. It is important to notice that Miraflores, the Presidential Palace where
Chávez resides, is located in the dense district of Libertador. Figure 7.3 clearly
shows that there is no direct correlation between land values and accessibility to

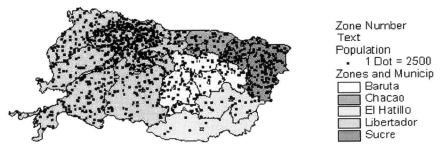

Figure 7.1. Municipalities and population in Caracas. The two most populous municipalities are Libertador and Sucre. Drawn by Josefina Florez and Nelliana Viloria in 2000 with data from 1998.

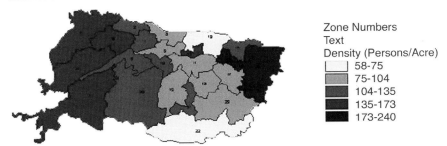

Figure 7.2. Density in Caracas. The densest areas are in the east and west of the metropolis, where squatter settlements and low to middle-low income housing areas are. The Presidential Palace where Chávez resides is located in the dense area of Libertador. Drawn by Josefina Florez and Nelliana Viloria in 2000 with 1998 data.

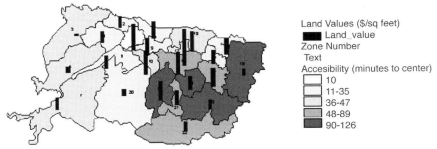

Figure 7.3. Relation between land values and accessibility to the centre of Caracas. Accessibility to the centre where the government institutions are located is measured in minutes of travel time. Elaborated by Josefina Florez and Nelliana Viloria in 2000 with 1998 data.

the centre of the city, where Miraflores and many of the government institutions are located.[10] The areas with highest land values are those with the lowest density and largest income, and correspond to formal urbanization. Alternatively, the areas with lowest land values have the highest density and lowest household income and mostly reflect informal urbanization. Figure 7.4 shows the distribution of yearly income per household. We can detect income polarization: most households in Libertador and Sucre have incomes up to US $6,300 per year as

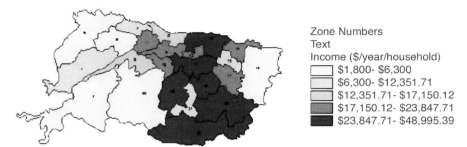

Zone Numbers
Text
Income ($/year/household)
☐ $1,800- $6,300
☐ $6,300- $12,351.71
☐ $12,351.71- $17,150.12
☐ $17,150.12- $23,847.71
■ $23,847.71- $48,995.39

Figure 7.4. Distribution of yearly income per household in Caracas. The map shows the income polarization that exists in the metropolis. Elaborated by Josefina Florez and Nelliana Viloria in 2000 with 1998 data.

of 1998, while households in Chacao, Baruta, and El Hatillo have incomes higher than $23,800 (except the *barrios* within Baruta where income is $12,300).

Figure 7.5 shows electoral results for Caracas of the 2004 referendum to end Chavez's presidency. The districts with largest population density and size and the lowest income and lowest land values voted to keep Chávez president. This geographical political distribution thus correlated with socio-economic, political-administrative and spatial factors and has had implications for the use of public space for political demonstrations, especially during April 2002, as discussed in the next section.

"¿Está usted de acuerdo con dejar sin efecto el mandato popular, otorgado mediante elecciones democráticas legítimas al ciudadano Hugo Rafael Chávez Frías, como presidente de la República Bolivariana de Venezuela para el actual periodo presidencial?"

Referendo1.dwg
■ NO
☐ SI

PARROQUIAS AREA METROPOLITANA DE CARACAS

Municipio Libertador
1. 23 de Enero
2. Altagracia
3. San José
4. San Bernardino
5. Catedral
6. Candelaria
7. Santa Teresa
8. San Agustín
9. El Recreo
10. San Pedro
11. Santa Rosalía
12. El Paraíso

13. San Juan
14. La Pastora
15. Sucre
16. El Junquito
17. Antímano
18. La Vega
19. El Valle
20. Coche
21. Caricuao
22. Macarao

Municipio Chacao
23. Chacao

Municipio Baruta
24. El Cafetal
25. Minas de Baruta
26. Nuestra Señora del Rosario de Baruta

Municipio El Hatillo
27. El Hatillo

Municipio Sucre
28. Leoncio Martínez
29. Petare
30. La Dolorita
31. Caucaguita
32. Filas de Mariche

Figure 7.5. The 2004 recall referendum results for Caracas. Elaborated by Nelliana Viloria, 2004.

Chronology of Political Events in Public Space

This section analyses the political and spatial dimensions of the events prior to and during the April 2002 *coup d'état* that forced President Chávez out of office for 43 hours. Significantly, people from many of the sectors that support Chávez in the west of the metropolis – as evidenced in the recent referendum (and in previous and subsequent elections) – have better accessibility to the city centre than people in the sectors that voted against him (figure 7.3). It is important to note, however, that Chávez supporters also hail from other sectors of the city (e.g. Petare, El Valle, and Caricuao) that do not have similar accessibility. Also not all *barrios*, although close, have 'good' accessibility. In other words, although accessibility definitely helped, even more important was the people's determination to restore Chávez to the presidency. This allowed significant numbers to be present around the Presidential Palace for the days leading up to the coup on 11 April 2002. After Chávez was abducted, supporters of the Chávez government quickly organized and by the next evening had congregated in large groups around the Palace and the Tiuna military fort. This presence and the support of a sector of the military which remained loyal to Chávez derailed the coup.

Precedents

In describing the political process that led to the April 2002 events, we adopt the national government's denomination of two distinct periods in recent Venezuelan history: the IV Republic for the period beginning with the fall of the last dictatorship in Venezuela in 1958 until 1998–1999, and the V Republic for the period following the election of President Hugo Chávez in 1998 and the subsequent promulgation of a new national constitution in 1999.[11] In each phase, the socio-political construction and use of space have been factors in the definition of the political process.

IV Republic. From the overthrow of the last dictator, Marcos Pérez Jiménez, in 1958 until the election of Chávez as president in 1998, Venezuelans elected presidents every five years, usually from the Social Democratic AD (*Acción Democrática*) or the Social Christian COPEI (*Comité de Organización Política Electoral Independiente*) parties. Public events during the campaigns and those for Labour Day and the yearly celebrations of party anniversaries were mostly orchestrated and left little room for spontaneous political demonstrations. The last were mainly restricted to student protests, sometimes supported by some faculty and other university or high school staff and members of smaller political parties.

Protests followed a repetitive pattern, often starting peacefully and joyously, but ending in violent confrontations with the police and sometimes injury and death. In this period, the sites of struggle were mainly restricted to spaces around public high schools (*liceos*), colleges, and universities. The principal public meetings were limited to the main plazas in the traditional centre of Caracas – Plazas O'Leary and Venezuela – or Bolívar Avenue (closed to traffic for the events). Regular protests in 'hot spots' (such as the Plazas Venezuela and *Las Tres Gracias,* both adjacent to the Central University of Venezuela; and José Antonio Paéz Avenue in front of the Pedagogic University) and the resulting traffic jams became predictable rituals. Probably due to the economic crisis, they subsided during the 1980s as less funds were available and because of the deterioration in public safety and disenchantment with traditional political parties.

The *El Caracazo* or *Sacudón* of 1989, a social reaction against the neo-liberal policies introduced by the penultimate president of the IV Republic, Carlos Andrés Perez, and in particular to the increase in public transport tariffs, signalled a break with established rituals. On 27 February, the poor blocked principal avenues and freeways and paralysed the city. The middle- and upper-income classes reacted with fear as their ever-present nightmare seemed to come true – i.e., the poor descending from their *barrios* on the hills and taking over the city. Expressing frustration, disenfranchised Caraqueños stormed and looted commercial centres and small businesses in a seemingly unorganized and anarchic form. In a brutal clampdown the police and the military killed an unknown number of people – a conservative estimate is 400, but other unofficial estimates place the number at 3,000.[12] Curfews and blockades lasted almost a week and represented a rapidly escalating crisis of representative democracy in the country. After El Caracazo, the challenges to the political establishment became more frequent and prominent. Street protests increased dramatically in Caracas, although some years were more turbulent than others. Subsequently, the perception of Venezuelan social history, as Lopez Maya and Lander explain, changed:

In the early 1980s, Venezuelans were thought of as being one of the least politically mobilized people of Latin America. It was argued that its solid democracy, oiled by the oil rent of the State, had allowed it to establish and to consolidate channels of efficient mediation and representation that kept pronounced social and/or violent conflict at bay… This has been contradicted in the last two decades and has forced the re-examination of 'street politics' in Venezuela. (Lopez-Maya and Lander, 2004, p. 1, authors' translation)

The political discontent of the 1980s and 1990s resulted from an acute economic crisis with increasing poverty and socio-economic polarization, rampant corruption, loss of political party legitimacy, lack of accountability and justice, and growing national debt. In addition, another element that sparked off the challenge to the political establishment after El Caracazo was the discontent of middle-level officers who felt frustrated with the country's political leaders and repugnance at having to take up arms against citizens during the riots, all culminating in two unsuccessful *coups d'état* against the government of Carlos Andrés Pérez in February and November 1992. The former was led by Colonel Hugo Chávez accompanied by other military leaders. From this moment on, a new political movement which had been secretly gestating as a reaction to the political crisis in the country became visible and gathered momentum. In addition, those organizations and movements which were excluded from the Pact of Punto Fijo[13] came together, despite class differences, to achieve the goal of 'democratizing democracy' (García-Guadilla *et al.*, 2004, p. 11). For many Venezuelans, this movement represented an alternative political project that offered greater hope for reform. Finally Pérez was impeached and Rafael Caldera was subsequently elected president in 1993. Falling international oil prices did not help Caldera, and his term was characterized by political instability and general dissatisfaction with the government. Many felt that he betrayed his constituency when he turned to the International Monetary Fund for aid (*Ibid.*), all of which provoked passionate street demonstrations demanding accountability and justice.

V Republic. The alternative political model, represented by the political parties identified with the project led by Chávez, garnered ample popular support and in 1998 won the national election with 56 per cent of the vote. Chávez assumed the presidency of the country, soon renamed by popular vote Bolivarian Republic of Venezuela. His presidency represented an interruption in the domination of *AD* and *COPEI*. The approval of a new Constitution in 1999 with 70 per cent of the vote gave definition to the 'peaceful revolution', as government labelled it. Shortly afterwards, new presidential elections were held and Chávez was re-elected for six years with possibility of re-election, as established by the new Constitution. The constitutional reform established a participatory democracy with popular involvement in government decision-making, the valorization of human rights, and the recognition of the multiculturalism of Venezuela.

In addition, the regime emphasizes Latin American integration and openly questions neo-liberal principles and conditions imposed by world financial bodies. Taxes, import duties, and licenses were reinvigorated, the oil industry was renationalized and its management revamped.[14] Land reform (to legalize the occupation of some urban and rural properties and redistribute others), and fishing rights (to control drag-net fishing) ensued. It is generally accepted that this was proof that the Chávez government would encourage substantial reform, if not revolution, alienating in the process sectors of the middle- and upper-income classes. A particularly thorny issue was land reform which lead to the first business shut-down in December 2001 (Wilpert, 2003*a*).[15]

As a result of constitutional change and its implementation, some vested interests in the church, traditional political parties, the military, the judicial system, the government bureaucracy, state and private companies, academia, the arts, and the media have been challenged. 'In countries besieged by extreme inequalities', García-Guadilla *et al.* (2004, p. 20) explain, 'class cleavages create deep divisions making universal proposals difficult to articulate'. Not surprisingly, a strong and vociferous opposition emerged, albeit one that is internally fragmented but with access to substantial economic resources and which dominates the mass media. Initially this appeared as a vigorous democratic opposition, but on 11 April 2002, anti-government forces manoeuvred into a position from which a *coup d'état* was fomented.

The Events of April 2002

By April 2002, demonstrations and marches gained intensity as Chávez and leaders of the opposition encouraged their followers to take to the streets. Each group sought to demonstrate that it could garner ever greater numbers of participants in different plazas and avenues of the city generating constant controversy over the exact numbers involved. The geography of congregation reflected the lines of class, race, and political affiliation, all of which were highly correlated. Some spaces were even renamed (e.g., Plaza Francia became *Plaza de la Libertad* (Liberty Plaza)), and exclusively appropriated and guarded by certain groups. These reterritorializations increased existing social and political cleavages.

On 11 April 2002, the opposition marched against Chávez's government (figure 7.6). From a nondescript space in front of a building of the PDVSA (the state oil company) in Chuao renamed by the opposition group as Meritocracy Plaza,[16] the march proceeded peacefully along the Francisco Fajardo Freeway. Although the legally permitted route was entirely within the Chacao and Baruta municipalities, some leaders of the opposition persuaded the marchers to cross into Libertador towards the Presidential Palace. As the march reached the boundary of Libertador, a group of entrepreneurs and high-ranking officers planned a *coup d'état* which was supported by representatives of important civic organizations. They attempted to legitimize their sedition by the sheer number of marchers as they resorted to unconstitutional, violent means to depose the president (García-Guadilla *et al.*, 2004).

The political climate in the country had been very tense for months, and credible rumours about the possibility of a coup attempt had surfaced. Therefore, enthusiastic government supporters had surrounded the Presidential Palace for days in a pre-emptive move. When the opposition march reached the area near the Palace on 11 April, security forces managed to keep government and opposition supporters apart, but as a result of what appeared to be sniper gunfire

Figure 7.8. Plaza Altamira, taken over and renamed Liberty Plaza by the opposition, 2002.

Christmas and return it to the residents. After a Chavista march led by the then Vice-President José Vicente Rangel, the plaza was finally 'reconquered'.[25]

Throughout the crisis and the concomitant dispute over territory, both groups resorted to symbolic warfare, appropriating nationalist and religious symbols. Chávez and his followers adopted the colour red for berets and T-shirts, symbolizing the leftist 'revolution'. The opposition, on the other hand, used yellow, symbolizing light and hope for a future without Chávez; or black, in sign of mourning. For some, the symbolic use of black alternatively created links, purposely or not, to European fascist movements. Black could also be interpreted as indicative of the diversity of ideological positions in the opposition and also as symbolic of a reactive, rather than an assertive, movement. Massive marches of Chavistas have been called *la marea roja* (the red tide), and those of opponents, *el gusano de luz* (the light worm). In a country with a Catholic majority, Chávez speaks often of Christ as leading his revolution; whereas the opposition invokes the Virgin Mary to get rid of Chávez. Chávez refers intensely to the new Constitution and shows it in many public appearances. The opposition, having voted against it and lost, later invoked it too, for the purpose of legitimizing the recall referendum. Lastly, Simón Bolívar, historical father of the Republic, after being admired by all Venezuelans, has recently turned into a symbol of division: Chavistas venerate him and hold him as the main inspiration of their 'Bolivarian Revolution', while anti-Chavistas are torn about what they see as an unwelcome revision of national history (González Deluca, 2005; Romero, 2005; Suazo, 2005).

The National Strike of Late 2002 and Early 2003

The polarization continued after the coup, fuelling social imaginaries that constructed the 'other' as enemy (Lozada, 2004; Salas, 2004). PDVSA, *Fedecámaras*, and the Central Labor Union formed an opposition coalition and planned for a general national strike and oil company stoppage to weaken the government and force Chávez's resignation. García-Guadilla *et al.* (2004, p. 20) describes the conditions as follows:

> The 2002 *coup d'état* did not put an end to the exclusion of the 'other' from the normative models proposed by each faction. The General Strike of December 2002 that lasted until February of 2003 carried out by the opposition in defiance to the government is another example of the use of the existential struggle as a means of articulating difference. The main objective of the general strike, called 'Paro Cívico' [Civic Stoppage] by the opposition, was to force President Chávez to resign by disrupting the national oil industry. This general strike was called by the *Coordinadora Democrática*, an entity that represented the alliances established between political parties and social and economic organizations belonging to the opposition. Civil organizations in the opposition supported the strike despite the fact the government did not recognize it as legal and even threatened to fire those public employees participating in it. The threat materialized when more than 5,000 professionals working in the State owned Oil Company, Petroléos de Venezuela (PDVSA), and its research institute INTEVEP, were fired. During the two months that the 'Paro Cívico' lasted, there were numerous street mobilizations and violent confrontations. The social imaginaries of the opposition and the government fueled the conflict. The government labeled citizens participating in the strike as the 'enemy', the 'traitors', and the 'anti-patriotic'. The strike was declared 'illegal'. The opposition, however, described the strike as a 'Paro Cívico' and justified their actions appealing to Article 350 of the Constitution that legitimizes Civil Disobedience. The 'illegal strike' or 'Paro Cívico' was de-activated only after it proved to be unsuccessful.

Many in the managerial and professional classes joined the strike, while owners of medium and small enterprises and blue-collar workers chose not to. As a result, the strike was unevenly adhered to in the city. Many sectors in the east closed their businesses as streets emptied out, while most sectors in the west, and poorer *barrios* elsewhere remained bustling with people and activity. Wealthier businesses were more likely to oppose Chávez's regime and could absorb the cost of a strike. Most small businesses, even if they were not Chávez supporters, could not absorb those costs. At PDVSA, managers for the most part complied with the strike, while many blue-collar workers did not and by defying their bosses recovered the company for the government. Many analysts also referred to the strike as a media event, because of the media's skewed depiction of the activities in different spaces of the city during the strike (Kaiser, 2003).[26]

The business-owner strike of 2002–2003 and the stoppage of the oil company resulted in an undersupply of gas, a collapse of transportation, and deficits in the supply of food and other goods. Those most hurt could not afford to hoard large quantities of food nor had space to store them. The middle class was also affected, because of its consumption of products and services imported or denominated in dollars (Wilpert, 2003a, p. 106). Securing gasoline or gas became a daily concern. In the *barrios*, people formed queues each time gas cylinders (used for cooking) were promised. Wherever lines formed, the surrounding spaces were converted, temporarily, into locations for further political discussion. People also created elements of self-ordering – controlling the lines, sharing knowledge regarding the locations where gas or gasoline were available, and securing some distributional justice.

Those supporting the government occupied key spaces such as the PDVSA headquarters in La Campiña (in Libertador municipality) in a relatively hostile area around the fortress-like headquaters.[27] Those supporting the stoppage camped in the previously mentioned Meritocracy Plaza in Chuao. The experience of the entrepreneurs' stoppage generated a government reaction and reorganization. Urgent action was needed on the part of the government to avoid risks to the supply of fuel and food. The government's provision of basic subsistence to lower-income classes (the Mercal mission) was strategic in its effects, and it also was maintained and reproduced after the strike as a way of pre-empting or co-opting events by the opposition (e.g., major open market events held at the time of opposition marches).

The Opposition Resorts to Violence: the Guarimba

On 27 February 2004, the fifteenth anniversary of El Caracazo, the opposition marched to recall President Chávez. In Caracas, a meeting of the leaders of unaligned countries (the G15 group) was being held in the Teresa Carreño Theater. For security reasons, the march was refused permission to proceed down Libertador Avenue (the theatre is at the end of the avenue), and violent clashes with the National Guard ensued. In protest, some members of the middle-class went on a three-day rampage, the *guarimba*, sequestering substantial areas in the east of the city. *Guarimba* acts continued between February and March 2004 in Caracas and some other Venezuelan cities. The aim was violent disruption of public order to provoke repressive military intervention, delegitimize the government, and provoke international action. Conceived by Robert Alonso, a member of the rightist group *Bloque Democrático*, the goal was to create an anarchy that would paralyse the country. In its extreme forms, *guarimba* tactics included burning car tyres and rubbish in public streets, and using homemade explosives like Molotov cocktails and conventional arms. In effect, the media activated a 'mediatic *guarimba*' (according to writer Earle Herrera) by questioning the legitimacy of public authorities and generating a general ambiance of terror. Fortunately, the call for massive participation in the *guarimbas* failed. Both support for Chávez and the democratic leanings of Venezuelans, including many from the opposition, doomed it (García Danglades, 2004). The *guarimbas* were lifted because they were also extremely unpopular as they hampered free movement and littered neighbourhoods (García-Guadilla, 2005).

The Referendum in August 2004

In contrast to the antidemocratic coup attempt and the illegal strike, the opposition also made use of the democratic strategy of the recall election, which came as a result of street and political mobilizations in 2003 and 2004.[28] Not since 1998 had the opposition displayed greater democratic strength and organizational potential, which convinced many of the inevitability of victory. The social and spatial segregation of opposition forces reinforced this impression. If everybody they knew, talked to, and encountered in their daily lives shared their political antagonism toward the government, how could they have conceived of not winning the recall? The media openly opposed to Chávez also played a big role in reasserting these perceptions in their biased portrayal of public demonstrations, marches, and exit polls. Government supporters, similarly, rejected the conventional mass media and came to depend on the State TV and radio stations as well as a limited number of alternative publications.

Thus, the participation in extraordinary events in public spaces in Caracas and the treatment of those events by the media had a determining role in the creation of the disparate imagined communities that Venezuelans of different political allegiances constructed as idealizations of their city and nation. The media construction of the opposition's powers was not reflected in the subsequent recall election results which, as depicted in figure 7.5, reconfirmed Chávez in power (with 59 per cent of the votes) and restated the geopolitical segregation of the city. The legitimacy of the elections was confirmed by international observers, including the Organization of American States (OAS) and *Centro Carter* led by United States ex-president Carter. However, many members of the opposition, in part motivated by some of their political leaders, and influenced by their media-constructed visions, did not recognize what they claimed were fraudulent electoral results. Their reactions were felt in the streets soon after.

Discussion of Socio-Spatial Dynamics

In the tensions between the creation and contestation of new political projects, 'imagined communities' are redefined, as are new subjectivities of solidarity and belonging. The effect for public space is an unsettling of the traditional meanings of place. Post-traditional revisions of pre-established national history are performed through the validation and weaving in of subaltern stories. Also, the current socio-political condition in Venezuela presents an ever more salient post-glocal dimension, as the global and the local embed in each other in myriad and fluid ways. Today, Venezuela is a main point of reference for comparative case studies conducted by international analysts and scholars. Activists and agencies from around the world participate in Venezuelan politics and in the local politics of public space, many of whom participate in the local marches. In 2006, the World Social Forum was held in Caracas and the Venezuelan president travelled around the world promoting his Bolivarian message and lobbying for Latin American supranational cooperation treaties. Political tourism to the country is growing.

One profoundly liberating transformation has been the way many people, who previously had a hard time barely 'making life' in the country, have now become effective subjects capable of 'making history' (Flacks, 1988). Thus, perhaps to a scale similar only to that in Cuba, the current process in Venezuela stands out with respect to other case studies in this book in that it reveals the inverse dynamics to the Habermasian notion of the colonization of the 'lifeworld' by the 'system'. The 'system' in Venezuela is experiencing critical transformations of yet unknown consequences. As a matter of fact, there are significant disputes among Chavistas as to whether the transformations should be of a reformist or revolutionary nature (Ellner, 2005), as both the national and international opposition to the regime organize powerful challenges to such changes. Paying attention to these dynamics not only can help us understand the unfolding of democracy and citizenship, reforms and revolutions; but also guide us toward more peaceful and just ends.

It is still to be seen how the so-called third-wave democracies of Latin America are transformed by the present 'grand refuse'. It is still too early to tell whether the current street politics amount to more than a reaction to neo-liberal policies and disenchantment with previous national political regimes and parties, and can catalyse a vision for alternative socio-political projects. But as people take their collective grievances and hopes to the streets, political consciousness is transformed. As the Venezuelan political regime pressures for the advancement of post-neo-liberal policies and practices, the country becomes the referential crest of the tide that is potentially opening Latin America to a new era, raising hopeful expectations for many and profound fears for others. In this process, the limits of both the invited and invented arenas of citizenship are being continuously redrawn (Cornwall, 2002; Miraftab, 2004). Caraqueños and other Venezuelans have taken to the streets of their capital to both denounce the 'post-justice city' (Mitchell, 2001) and explore what the 'just city' and country can be (Fainstein, 2000). Caracas has thus become the 'contesting city' (Cantú Chapa, 2005) or 'insurgent city' (Holston and Appadurai, 1999) *par excellence* in Latin America and as *The New York Times* puts it, the 'new Mecca for the left' (Forero, 2006).

Cultural citizenship in Caracas has been collectively constructed and expanded since 1998 (Delanty, 2002). Citizenship has also been shaped through struggles for rights in the disputes over meanings and practices of participation, the distribution of resources, and boundary-setting between the public and the private, the social and the individual, and the modern and the traditional (Tamayo, 2004). The previously excluded and disenfranchised on the basis of race and class, and also sex, gender, religion, culture, and national origins have reclaimed their citizenship (Kearns, 1995; Miraftab, 1994). But though 'insurgent' (Holston, 1995), it does not necessarily amount to inclusive citizenship (Gaventa, 2002). As these struggles, most of which have been played out as zero-sum games, progressed, others have voluntarily opted out of the current process hoping that their non-participation will delegitimize the Chávez government.

Despite these challenges, inclusive citizenship in Caracas is being constructed as it is being actively performed, fully representing the dynamics of active and performative citizenship discussed by Kearns (1995), Miraftab (1994), and Boudreau (2000), respectively. Concretely, in the recent Venezuelan constitution, citizenship rights have been expanded for women, indigenous groups, and the dispossessed. A gap, however, remains between citizenship as defined by law and as it is enacted in everyday life, described as a 'performative paradox' (Lee, 1998). It is not widening because people challenge the law with their practices making it more inclusive, which is the most common case when a performative paradox is in place. Instead, the novelty of the current performative paradox in Venezuela is that it challenges individual and institutional practices – from all sides of the political spectrum – that do not live up to the expectations created by the new constitution and derived laws. The extraordinary, short-lived events in Caracas's public places continue to have a transformational effect in the city and the nation. These spaces of insurgent citizenship 'engage, in practice, the problematic nature of belonging to society' (Holston and Appadurai, 1999, p. 50), and explicitly challenge the post-justice city as the inequitable result of an unchecked neo-liberalism. For many, they offer hope for the future of the just city, yet its realization is not a given. The difficult trajectory from raising political consciousness to effective political action on progressive policy remains to be traversed. The hurdles are both external, i.e., the national and international opposition to the regime, and internal, as the corruption, opportunism, ineptitude, and exclusionary politics of the IV Republic are not so easily surmounted in the V Republic.

Two of Winocur's (2003) perspectives on citizenship transformation (Chapter 1) are relevant here. Firstly, citizens have rights to quality public space. Accordingly, minority and marginalized groups should have rights to use public space as a strategy of visibility and survival. This has been accomplished in Caracas to a great extent by new practices of social interaction and control, consumption and production of space. Arguably, a sense of responsibilities associated with the use of the 'commons' have not developed at the same pace as notions of rights of use, and thus problems of maintenance and preservation have in some cases worsened.

The second perspective explores the impact of the media in transforming participation in the public sphere and in public spaces, which we have shown to be extensive. The disingenuous portrayal of demonstrations, marches, stoppages, strikes, riots, and exit polls have all lead to 'imagined communities' (Anderson, 1983) and 'imagined geographies' (Gregory, 1994) in Venezuela. These are built on people believing that the city they inhabit is the one that is distortedly re-presented to them through selected and edited media images and discourses. Conversely, the media has been cleverly and effectively used to secure and expand spaces and practices of insurgent citizenship. In effect, the hegemonic role of the media resulted in an informational 'grand refuse' as 'grass-roots' communication (mouth to mouth, computer to computer, and cell phone to cell phone) reversed the 2002 *coup-d'état*.

Thus, the challenge remains for further *radicalization of democracy* in Venezuela. This radicalization of democracy has progressed through the repeated legitimation of this regime through elections, the enactment of a new constitution and laws, and extending the rights to street politics for all Venezuelans. This, however, has been hindered by the aforementioned corruption, opportunism, ineptitude, and exclusionary politics on the part of the government and non-cooperation and outright sabotaging on the part of national and international opponent elites. To enable a state of full participation and deliberation, respect for both supporters and dissenters of the new political regime is necessary.

Final Words

In the complex and ongoing street politics in Caracas discussed in this chapter, there is a range of reterritorializing and deterritorializing processes at play. From the point of view of urban space, several points should be remembered. First, public spaces in Caracas have been the scenes for expressing, both peacefully and violently, a wide range of political stances both for and against the new regime. During the coup and its aftermath, public spaces were critical sites for both the enactment of discontent and the reformulation of citizenship through marching, chanting, rioting, graffiti, and other street acts, drawn from both traditional and more insurrectional tactics. Second, beyond the incidents taking place in real-time/real-space, a critical, superposed dimension to the fatidic April 2002 events and its aftermath can be identified. It is constituted by the mismatch between those real and variegated happenings occurring in different areas of metropolitan Caracas, and the virtual city presented simultaneously by the media through the manipulated construction and editing of partial truths. In an attempt to counter manipulated media information, public spaces have been disputed and occupied by the conflicting groups and used as expressions of solidarity and face-to-face communication among members of the same political affiliation.

From the point of view of architecture, public buildings – seats of political and military power in Venezuela – on the one hand and private buildings – headquarters of the media, the oil company, and others in the opposition – on the other, functioned as symbolic icons of power around which the public has gathered. Significantly, both the metropolitan location of these buildings in Caracas and their accessibility from different parts of the city have been crucial.

This study reveals that in Venezuela the challenge remains for political actors to find ways 'to transform the existential struggle back into a dispute' (García-Guadilla *et al.*, 2004, p. 20). On the one hand, the state has to regain legitimacy in the eyes of all Venezuelans and to establish bridges to all sectors to negotiate conflict through constitutional means. On the other hand, civil society should strive to define a common collective identity to relate to the state and recognize its mediating role. Urban spaces and buildings have become symbolically and practically imbricated with the ongoing conflicts, enabling social inequalities and polarization in the country to find their expression and contestation in the capital's urban space. Just as extraordinary events in Caracas's recent past have marked the current state of events in Venezuela, ordinary places can be catalysts for citizen awareness to manoeuvre from undemocratic existential struggles to a democratic and participatory political culture. In distinguishing between making life and making history (Flacks, 1988), we celebrate that people in Caracas have risen to the challenge of making history through extraordinary political mobilization in public space. We hope that they also seize this unique opportunity to refashion ordinary life into being democratic, inclusionary, and just.

Notes

1. The literature on this is extensive. For syntheses, see Ellner and Hellinger (2004); McCaughan (2004).
2. Exceptional contributions in this area have been done by García-Guadilla (1998; 2002; 2003; 2004); García-Guadilla *et al.* (2004), López Maya (1999, 2003), López-Maya and Lander, (2004), Lander and López-Maya (2005), Navarrete (2005).
3. According to the *Oil and Gas Journal* and Energy Information Administration, there were eight countries with greater annual oil production than Venezuela, but only five with greater proven reserves, all being in the Middle East (Appenzeller, 2004, pp. 90–91). This underlines the strategic importance of Venezuela to the United States, which is by far the greatest consumer of oil in the world, followed by China and Japan. 13 per cent of US oil imports come from Venezuela (*Ibid.*, p. 89).

4.   This approximation belies the difficulty of defining the limits, density, and urban form of self-built squatter settlements which are estimated to house 60 per cent of Venezuela's population (Architectural League of New York, 2002).

5.   The population of Caracas could be larger because figures tend to underestimate the *barrio* population because of the difficulties of access and of accounting for non-citizen residents. The most populous district, Libertador, had a population of 1.9 million in 2004.

6.   This public housing project is composed of 70 buildings with a total of 9,176 apartments, and some community services (schools, markets, etc.). It is named to mark the outing of the last dictator Pérez Jiménez in 1958. Originally planned for 60,000 inhabitants, today it is home to 76,000 people. Counting the self-built houses between buildings, the area's total population is in the hundreds of thousands. The area population is particularly politicized, and it is within walking distance of the Miraflores presidential palace.

7.   These municipalities were created by the division of the old Sucre District in the municipal reforms of 1989.

8.   There are some important disparities in the statistics found for this period. In this quote, the percentage of poor are different from the figures given above. Additionally, the figure of 60 per cent middle class in 1982 more or less corresponds to the 'not poor' category in other accounts, but this category is not necessarily equivalent to 'middle class'.

9.   All the maps were elaborated by Josefina Florez and Nelliana Viloria in 2000 with information updated for 1998. These maps were updated from originals from 1992, which formed part of Josefina Florez's PhD dissertation 'Accesibilidad, calidad urbana y grupos socioeconómicos en el patrón de localización residencial' (Spain, 1992). The primary sources of information were: Income: stratification proposed by Datanálisis for socio-economic sectors from A to E, 1998. Average value of the US dollar: Banco Central de Venezuela. Density: OCEI projections for 1998. Land use: AKROS registries from January to December 1998, following a methodology by Rivas and Padrón. Accessibility: Josefina Florez.

10. It is important to bear in mind that Caracas is multinucleated, it has a rapid transportation system (the Metro), and the majority of people in poor areas do not own autos, all of which affects land values.

11. The constitution of 1811 marked the beginning of Venezuela's I Republic, but since the Caracas *criollo* elite failed to rally popular support for the cause of independence, a racially defined civil war marred the early years. In 1813, Simón Bolívar captured Caracas, where he was proclaimed 'The Liberator' establishing the II Republic. But by 1814, enemy troops forced Bolívar and his army out of Caracas, bringing an end to the II Republic. Although Caracas remained in royalist hands, the 1819 Congress at Angostura (present-day Ciudad Bolívar) established the III Republic with Bolívar as president. In 1821, Bolívar's troops finally liberated Caracas from Spanish rule. Nearly a century and a half of military rule ensued until 1959. This stark break with the past has been attributed to oil-based wealth, which gave government the material resources to win the population over to democracy, and to the spirit of cooperation among the nation's various political entities as embodied in the Pact of Punto Fijo. In another 1958 pact, the 'Declaration of Principles and Governing Program', the main political parties (AD, COPEI, and the URD) agreed on a range of matters with respect to the economy, including the rights of association and collective bargaining, and state subsidies for the poor. The objective was to institutionalize a 'prolonged political truce' in favour of the democratic project. The 'Spirit of the 23rd of January' informed the 1961 constitution, which guaranteed a wide range of civil liberties, instituted a weak bicameral legislature, and strong executive powers. The major group excluded from the political pacts of 1958 was the extreme left (Haggerty, 1993, np).

12. Owing to the discovery that the government had buried civilians in mass graves and not counted those deaths.

13. In the 1958 Pact of Punto Fijo, president Betancourt granted concessions to a broad range of political forces, but yet managed to maintain the hegemony of the main political

parties, especially AD and COPEI. Other groups were excluded from the political pacts, particularly the leftists (Haggerty, 1993, np). Chavistas conceived the Pact of Punto Fijo as an essential symbol of the IV Republic.

14. The new oil law 'limited foreign companies to 50 joint ventures and doubled the royalties to be paid to the state per barrel of extracted oil. It also for the first time imposed some accounting and fiscal transparency on the murky operations of PDVSA, and contained provisions allowing the government to restructure the petroleum industry in due course' (Wilpert, 2003a, p. 107).

15. See Wilpert (2003a); Vila Planes (2003); Camejo (2002); and Medina (2001) for a more detailed description of these reforms and their implications.

16. Chávez had claimed PDVSA's managers were committed to neo-liberal agendas which favoured foreign interests. Their status threatened, PDVSA's leading technocrats claimed they alone had the technical expertise necessary to run the company. In their defence, the space in front of the PDVSA building was renamed Meritocracy Plaza.

17. Rodríguez (2005, p. 191), who was attorney general (fiscal) at the time of the coup, gives the following names of the sharpshooters: Roberto Francisco McKnight, Roger de Jesús Lugo Miquilena, Franklin Manuel Rodríguez, Jorge Hernán Meneses, Jesús Antonio Meneses, Nelson Enrique Rosales and John Carlos Garzón. 'This last person was Colombian, another one was from Panama, McKnight was from the US and Jesús Lugo Miquilena had a false Venezuelan ID' (our translation). Not all those killed were demonstrators, some local workers were caught in the fray too.

18. What the commercial television channels did not show was the street below the Llaguno Bridge, which was empty except for the Metropolitan Police. The manipulation of information is evident in the documentaries *Puente Llaguno, Clave de una masacre* (Palacios, 2004b) and *The Revolution Will Not Be Televised* (Bartley and O'Briain, 2003). *Puente Llaguno* obtained the first prize in the Latin American Documentary Festival 2004, and the General Public award in Spain's Film Festival 2005. *The Revolution Will Not Be Televised* has also won numerous international awards.

19. Given these circumstances, one of the most notable aspects of the organization of the demonstrations in favour of Chávez was the capacity to communicate by alternative means. This confirms the importance of face to face contacts in public spaces – locally called *radio 'bemba'* (popular term for mouth, referring to oral communication).

20. This is so considering that government supported TV and radio stations were closed down by coup sympathizers, and that commercial channels were showing only comics and feature films. The only information some Venezuelans had access to was that broadcast by international cable channels such as CNN.

21. The Metropolitan police played a major role, encouraged by the then Metropolitan Mayor Peña who supported the coup. By this time, however, large factions of the military forces had declared their loyalty to the Constitution and the democratically elected government, making systematic violence difficult to maintain or to justify. Meanwhile, an involvement of the US government in the coup had been claimed, and the US did publicly recognize Carmona's government. We could suppose that the US government dropped its official approval later because it would have been difficult to justify support for a regime that had adopted measures that seriously violated human rights.

22. An NGO of coup victims (ASOVIC) lists 19 people killed on 11 April, and 27 between the 12th and the 14th.

23. The Cuban Embassy was attacked because it symbolically represents the communist ideology many opponents of Chávez claim is embraced by his regime.

24. The mayor of Baruta, Henrique Capriles Radonski, was charged in relation to his participation in the violent Cuban Embassy siege in Caracas between 9 and 13 April 2002. The main evidence against him can be found in celebratory footage aired by commercial TV. See the documentary *Asedio de una Embajada* (Palacios, 2004a).

25. This plaza continues to symbolize opposition to the government and is always a reference in the spasmodic protests against the government, as happened during the time of the *guarimbas*.

26. For instance, TV channels selectively showed streets in the anti-Chavista areas of the city where businesses were closed, denoting compliance with the stoppage.
27. This building is typical of the modern, fortress style public buildings of the IV Republic, which erected real and symbolic barriers between the public, especially the poor sectors, and the officials who were supposed to be serving them.
28. The 1999 Constitution provides for a recall of elected representatives after they have served half their term. It is claimed that the Venezuelan Constitution is the only one in the world that sanctions a presidential recall.

References

Anderson, Benedict (1983) *Imagined Communities: Reflections on the Origin and Spread of Nationalism.* London: New York, NY: Verso.
Appenzeller, Tim (2004) The end of cheap oil. *National Geographic*, **205**(6), pp. 80–109.
Architectural League of New York (2002) Barrio: No plan as plan. http://www.world view.org/caracas/barrios.html. Accessed 27 February 2006.
Bartley, Kim and O'Briain, Donnacha (dir.) (2003) *La Revolución no será transmitida (The Revolution will not be televised).* Documentary. Falvey, Sarah (prod.). Irlanda, 74 min: son., col.
Boudreau, Julie-Anne (2000) *The Megacity Saga: Democracy and Citizenship in this Global Age.* Montreal: Black Rose Books.
Caldeira, Teresa (2000) *City of Walls: Crime, Segregation, and Citizenship in São Paulo.* Berkeley, CA: University of California Press.
Camejo, Yrayma (2002) Estado y mercado en el proyecto nacional-popular bolivariano. *Revista Venezolana de Economía y Ciencias Sociales*, **8**(3), pp. 13–39.
Cantú Chapa, Rubén (2005) *Globalización y centro histórico: Ciudad de México, medio ambiente sociourbano.* Mexico: Plaza y Valdés.
Cornwall, Andrea (2002) Locating citizen participation. *IDS Bulletin*, **33**(2), pp. 49–58.
Delanty, Gerard (2002) Two conceptions of cultural citizenship: a review of recent literature on culture and citizenship. *The Global Review of Ethnopolitics*, **1**(3), pp. 60–66.
Ellner, Steve and Hellinger, Daniel (eds.) (2004) *Venezuelan Politics in the Chávez Era: Class, Polarization, and Conflict.* Boulder, CO: Lynne Rienner.
Ellner, Steve (2005) Revolutionary and non-revolutionary paths of radical populism: Directions of the *Chavista* movement in Venezuela. *Science & Society*, **69**(2), pp. 160–190.
Fainstein, Susan (2000) New directions in planning theory. *Urban Affairs Review*, **35**(4), pp. 451–478.
Flacks, Richard (1988) *Making History vs. Making Life.* New York, NY: Columbia University Press.
Forero, Juan (2006) Caracas calling: A new mecca for the left. *The New York Times*, Wednesday, 22 March. http://www.iht.com/bin/print_ipub.php?file=/articles/2006/03/21/news/venezuela.php.
García Danglades, Antonio Guillermo (2004) La 'guarimba', delito común y crimen contra la paz. *Aporrea*, 3 November. http://www.aporrea.org/actualidad/a7337.html.
García-Guadilla, María Pilar (1998) Ajuste económico, desdemocratización y procesos de privatización de los espacios públicos en Venezuela. *Revista Interamericana de Planificación*, **30**(119/120), pp. 77–89.
García-Guadilla, María Pilar (2002) Actores, organizaciones y movimientos sociales en la Venezuela del 2002, in Ramos, Marisa R. (ed.) *Venezuela: rupturas y continuidades del sistema político (1999–2001).* Salamanca: Ediciones Universidad de Salamanca.
García-Guadilla, María Pilar (2003) Territorialización de los conflictos sociopolíticos en una ciudad sitiada: ghettos y feudos en Caracas. *Revista Ciudad y Territorio: Estudios Territoriales*, Nos 136–137, pp. 421–440.
García-Guadilla, María Pilar (2004) Civil society: institutionalization, fragmentation, autonomy, in Ellner, Steve and Hellinger, Daniel (eds.) *Venezuelan Politics in the Chávez Era: Class, Polarization, and Conflict.* Boulder, CO: Lynne Rienner, pp. 179–196.

García-Guadilla, María Pilar (2005) The democratization of democracy and social organizations of the opposition: theoretical certainties, myths, and praxis. *Latin American Perspectives*, Issue 141, Vol. **32**(2), pp. 109–123.

García-Guadilla, Maria Pilar, Mallen, Ana and Guillén, Maryluz (2004) The Multiple Faces of Venezuelan Civil Society: Politization and its Impact on Democratization. Proceedings of XXV LASA (Latin American Studies Association) International Congress, Las Vegas, 7–9 October.

Gaventa, John (2002) Exploring citizenship, participation and accountability. *IDS Bulletin*, **33**(2), pp. 1–11.

González Deluca, María Elena (2005) Historia, usos, mitos, demonios y magia revolucionaria. *Revista Venezolana de Economía y Ciencias Sociales*, **11**(2), pp. 159–186.

Gregory, Derek (1994) *Geographical Imagination*. Oxford: Blackwell.

Haggerty, Richard A. (ed.) (1993) *Venezuela: A Country Study*, 4th ed. Washington, DC: Federal Research Division, Library of Congress. http://hdl.loc.gov/loc.gdc/cntrystd,ve. Accessed 26 February 2006.

Holston, James (1995) Spaces of insurgent citizenship. *Planning Theory*, **13**, pp. 35–52.

Holston, James and Appadurai, Arjun (1999) Cities and citizenship, in Holston, James (ed.) *Cities and Citizenship*. Durham, NC: Duke University Press.

Kaiser, Patricia (2003) Estrategias discursivas antichavistas de las medios de comunicación. *Revista Venezolana de Economía y Ciencias Sociales*, **9**(3), pp. 231–253.

Kearns, A. (1995) Active citizenship and local governance: political and geographical dimensions. *Political Geography*, **14**, pp. 155–175.

Lander, Luis and López-Maya, Margarita (2005) Referendo revocatorio y elecciones regionales en Venezuela: Geografía electoral de la polarización. *Revista Venezolana de Economía y Ciencias Sociales*, **11**(1), pp. 43–58.

Lee, Benjamin (1998) Peoples and publics. *Public Culture*, **10**(2), pp. 371–394.

López-Maya, Margarita (1999) La protesta popular venezolana entre 1989 y 1993 en el umbral del neoliberalismo, in López-Maya, Margarita (ed.) *Lucha popular, democracia, neoliberalismo: protesta popular en América Latina en los años del ajuste*. Caracas: Nueva Sociedad, pp. 211–235.

López-Maya, Margarita (2003) Movilización, institucionalidad y legitimidad. *Revista Venezolana de Economía y Ciencias Sociales*, **9**(1), pp. 211–228.

López-Maya, Margarita and Lander, Luis (2004) Novedades y continuidades de la protesta popular en Venezuela. Proceedings of XXV LASA (Latin American Studies Association) International Congress, Las Vegas, 7–9 October.

Low, Setha (2000) *On the Plaza: The Politics of Public Space and Culture*. Austin, TX: University of Texas Press.

Low, Setha and Smith, Neil (eds.) (2006) *The Politics of Public Space*. Oxon: Routledge.

Lozada, Mireya (2004) El otro es el enemigo: Imaginarios sociales y polarización. *Revista Venezolana de Economía y Ciencias Sociales*, **10**(2), pp. 195–209.

McCaughan, Michael (2004) *The Battle of Venezuela*. London: Latin America Bureau.

Márquez, Patricia and Piñango, Ramón (2003) *En esta Venezuela: Realidades y nuevos caminos*. Caracas: Ediciones IESA.

Medina, Medófilo (2001) Chávez y la globalización. *Revista Venezolana de Economía y Ciencias Sociales*, **7**(2), pp. 115–128.

Méndez, Miguel (2004) *Conflicto y reconciliación en Venezuela*. Caracas: Alfadil Edicones.

Miraftab, Faranak (2004) Invented and invited spaces of participation: Neoliberal citizenship and feminists' expanded notion of politics. *Journal of Transnational Women's and Gender Studies*, **1**(1). http://web.cortland.edu/wagadu/vol1-1toc.html (accessed December 2005).

Mitchell, Don (2001) Postmodern geographical praxis? The postmodern impulse and the war against the homeless in the post-justice city, in Claudio Minca (ed.), *Postmodern Geography: Theory and Praxis*. Oxford: Blackwell, pp. 57–92.

Navarrete, Rodrigo (2005) Presentación: ¡El pasado está en la calle! *Revista Venezolana de Economía y Ciencias Sociales*, **11**(2), pp. 127–140.

Palacios, Angel (dir.) (2004*a*) *Asedio a una Embajada* (Panafilms Documentary). Venezuela: Asociación Nacional de Medios Comunitarios, Libres y Alternativos.

Palacios, Angel (dir.) (2004b) *Puente Llaguno, Clave de una Masacre* (Panafilms Documentary). Venezuela: Asociación Nacional de Medios Comunitarios, Libres y Alternativos.

República Bolivariana de Venezuela. Instituto Nacional de Estadísticas (2001) *2001 Census*. Caracas: INE.

Riutort, Matias (1999) El Costo de Eradicar la Pobreza, in *Un Mal Posible de Superar*, Vol. 1, Caracas: Poverty Project of the Catholic University Andres Bello UCAB.

Roberts, Kenneth (2004) Social polarization and the populist resurgence in Venezuela, in Ellner, Steve and Hellinger, Daniel (eds.) *Venezuelan Politics in the Chávez Era: Class, Polarization, and Conflict*. Boulder, CO: Lynne Rienner, pp. 55–72.

Rodríguez, Isaías (2005) *Abril comienza en octubre*. Caracas: no publisher given.

Romero, Juan Eduardo (2005) Usos e interpretaciones de la historia de Venezuela en el pensamiento de Hugo Chávez. *Revista Venezolana de Economía y Ciencias Sociales*, **11**(2), pp. 211–236.

Salas, Yolanda (2004) 'La Revolución Bolivariana' y 'La Sociedad Civil': La construcción de subjetividades nacionales en situación de conflicto. *Revista Venezolana de Economía y Ciencias Sociales*, **10**(2), pp. 91–109.

Santamaría, Gema (2004) El trágico triunfo de Chávez, o la democracia que nunca fue. *Foreign Affairs en Español*, **4**(4), pp. 70–78.

Sassen, Saskia (1996) Whose city is it? *Traditional Dwellings and Settlements Review*, **8**(10), p. 11.

Suazo, Félix (2005) Usos politicos de la memoria: Devoción, desdén y asedio de las estatuas. *Revista Venezolana de Economía y Ciencias Sociales*, **11**(2), pp. 251–258.

Tamayo, Sergio (2004) Espacios ciudadanos, in Rodríguez, Ariel and Sergio Tamayo (eds.) *Los últimos cien años, los próximos cien*. México DF: Universidad Autónoma Metropolitana.

UN/CEPAL (2001) 57. [Report] Foro sobre el delito y la sociedad. Vol.1, No.1, February. NewYork: Naciones Unidas, Centro para la Prevención International del Delito.

Vila Planes, Enrique (2003) La *economía social* en el proceso bolivariano: Ideas controversiales. *Revista Venezolana de Economía y Ciencias Sociales*, **9**(3), pp. 111–143.

Wilpert, Gregory (2003a) Collision in Venezuela. *New Left Review*, **21**(May/June), pp. 101–116.

Wilpert, Gregory (2003b) Mission impossible? Venezuela's mission to fight poverty. *Venezuelan Views, News, and Analysis.* 11 November. www.venezuelanalysis.com/articles.php?artno=1051. Accessed 27 February 2006.

Winocur, Rosalía (2003) La invención mediática de la ciudadanía, Ramírez Kuri, Patricia (ed.) *Espacio público y reconstrucción de ciudadanía*. Mexico: FLACSO.

Chapter 8

The Struggle for Urban Territories: Human Rights Activists in Buenos Aires

Susana Kaiser

Every Thursday at 3.30 pm in the Plaza de Mayo of Buenos Aires, a group of women wearing white scarves holds a silent march circling the Pyramid of May, the central monument of the square. They are accompanied by supporters, watched by the curious, and ignored by rushing passers-by, but almost always monitored by police. For half an hour, the cause of the *desaparecidos* (disappeared people), the cry against repression, and the quest for justice are embodied in a public demonstration that activates society's collective memory. The march is a ritual of both resistance and commemoration performed at a highly symbolic location, and is one of the most intriguing acts of memory about the Argentine dictatorship (1976–1983). During this era, in a programme of state terrorism aimed to eliminate political dissent, the military used kidnap, torture, killing, and disappearances, and left a legacy of an estimated 30,000 *desaparecidos*. Since their irruption into the public sphere in the midst of the repression in April 1977, the Mothers of the Plaza de Mayo (mothers of *desaparecidos*) pioneered the redefinition of what is public that is at the core of human rights struggles in the country. By turning motherhood into a public activity, they were crucial in resetting the boundaries of politics and political spaces. By conquering and remapping territories, both physical and metaphorical, they shaped the style and scope of human rights activism.

In this chapter, I argue that in order to analyse human rights struggles in Argentina we cannot disregard the forms of participation and debates that result when people take to the streets. By exploring the human rights movement's struggles for urban territories over the last three decades, I discuss the key roles that the creative, strong, noisy, and disruptive public presence of activists had in shaping public opinion and policies regarding memory, accountability, social justice, and democratization. I start with an overview of the Mothers of the Plaza de Mayo's communication strategies in denouncing state terrorism and demanding accountability. Then, I analyse the *escraches* (demonstrations) organized by HIJOS (the organization of children of *desaparecidos*). The section on *escraches* forms the core of the chapter, where I discuss this tool as an innovative challenge to impunity and political amnesia.[1] Finally, I focus briefly on street demonstrations (2001–2002) as tools to bring down the government and as methods to demand major restructuring of political institutions and the economy. These demonstrations include *cacerolazos* (banging saucepans), *piquetes* (blocking of roads) and, in particular, 'new *escraches*', demonstrating how this strategy of denunciation has been co-opted and adapted to bring to public attention issues ranging from political corruption to biased media coverage of events.

The events discussed here took place during distinct and changing political environments. The Mothers' activism began during a military regime whose reign of terror had curtailed most signs of public resistance. When the dictatorship ended in 1983 the democratization process had a promising beginning. There was a commission to investigate disappearances and, in 1985, the military juntas were put on trial. But the promise soon faded as steps to legalize impunity marked the years of civilian rule. Authorities enacted impunity laws and granted pardons to the few *represores* (a generic term for torturers, assassins, and their accomplices) who actually ended up in prison.[2] In the light of these policies, the 1990s witnessed continued action on the part of activists who refused to give up their right to justice and denounced the politicians' unfulfilled promises. Finally in 2003, there was a major switch regarding accountability. Congress nullified the impunity laws, which resulted in the reopening of several cases against *represores*.

The legacy of the dictatorship is not the only human rights issue creating turmoil and prompting activism today. Deteriorating economic conditions have resulted in the denial of basic human rights to large segments of the population. Ironically, in the aftermath of the recent economic crisis, many Argentinians can vote but cannot afford a meal. The rates of unemployment and underemployment are alarmingly high, and all the while large swaths of the population are barely surviving below the poverty line. The devastating effects of the neoliberal economic policies initiated during the dictatorship and fully implemented during the 1990s, in combination with rampant corruption and unsolved crimes, among other 'ingredients', have created a political environment reeling under layers of problems.[3] At times, these override those attributable to the dictatorship, while at others they appear as 'aftershocks' of that sinister past.

Society does not remain passive though, and people organize, protest, and demonstrate. The streets of Buenos Aires are the places where denunciation, debate, contestation, and negotiation take place, the sites to voice demands for truth, accountability, and social justice. They are veritable 'arenas of power in which common action is coordinated' (Benhabib, 1992, p. 78).[4] As Ryan (1992) notes, those who are more removed from formal institutions find other means to make their grievances heard, including louder voices, civil disobedience, and violence. And we must not disregard the 'loudness' of public silent marches and vigils. Thus, public demonstrations and taking to the streets are examples of alternative media strategies that have been employed by social and political movements to advance a variety of causes and objectives.[5] The strong street presence that characterizes the Argentine human rights movement ranges from marches to press for accountability to unemployed workers blocking highways and demanding jobs. These forms of activism have created alternative public spheres, triggered changes, and influenced policy. They throw into question Habermas's (1979, p. 200) claim that laws resulting from the 'pressure of the street' could not be understood as 'arising from the consensus of private individuals engaged in public discussion'.

Moreover, taking to the streets is aimed at conquering and remapping cultural, political, and ideological zones that, as González (1999) reminds us, are never void but loaded with presences, symbols, and meanings. Since memory is encoded in places and takes root in spaces, the struggle for urban territories is thus also a means of enacting memory and writing history. Theorizing about the spatial frameworks of collective memories, Halbwachs (1992) argued that people only remember what is linked to localizable spaces. The stability of the spatial image is what gives the illusion of immutability over time and the possibility of recapturing the past in the present. Here, Halbwachs refers to physical locations but also includes legal, economic, and religious spaces, which are not always tangible as is the case with the deliberations within a court of law.

This notion links well with Nora's concept of the *lieux de mémoire* (memory places), which he uses to discuss those symbolic sites that codify memory as self-referential signs that have a capacity to recycle their meaning endlessly (such as geographical sites, historical figures, or monuments) (Nora, 1989, p. 9). The breadth of Nora's concept includes non-static and portable sites like the Torah, transitory and ephemeral sites like veterans' organizations, and even a more intangible application such as a historical generation – for example, the one of 1968 in Paris. Nora also distinguishes between 'dominant' and 'dominated' sites; the former are generally celebrations, officially imposed from above, to which one is summoned – i.e., de Gaulle's funeral at Notre Dame; the latter are 'places of refuge, sanctuaries of instinctive devotion' – i.e., Jean Paul Sartre's funeral (Nora, 1996, p. 19). Thus, we can say that Pablo Neruda's funeral during the terror of

Pinochet's Chile was an act of memory and resistance, and became a dominated *lieu de mémoire*. Again in Argentina, conflicts around dominant and dominated sites characterize official and unofficial politics of memory about the dictatorship.

Recognizing the value of intangible sites, and broadening the scope of what was identified as World Heritage, in 2003 UNESCO adopted the Intangible Cultural Heritage Convention. This category focuses on living cultural heritage such as oral traditions, social practices, rituals and festive events, languages, or performing arts that 'transmitted from generation to generation, is constantly recreated by communities and groups, in response to their environment and their historical conditions of existence' (http://portal.unesco.org/culture). There is a strong connotation of identity and memory in this category, which includes the carnival of Barranquilla in Colombia and the Indigenous Festivity Dedicated to the Dead in Mexico. The idea that an event that takes place in the same physical space might be different day to day and year to year (for example, a carnival) relates to the issues discussed here. Thus, we could suggest the notion of rotating *lieux de mémoire*, embodying with meaning particular ordinary spaces on an ephemeral basis and belonging to a sub-category of memory-centred intangible heritages. And we may categorize the Mothers' marches and the HIJOS's *escraches* as memory-centred intangible heritages and dominated *lieux de mémoire*.

In this sense, I also want to consider the geography of memory and the role that social actors play in loading spaces with new meaning. Local geography is a structure for remembrance that organizes the manner in which historical referents are conceptualized. Both in urban and rural landscapes people locate their memories within spaces such as squares, houses, factories, or caves (Fentress and Wickham, 1992; Rappaport, 1990). How can this help us understand the urban landscapes as experienced and transformed during the terror? How do people locate in these landscapes memories of traumatic events? In the Buenos Aires of the dictatorship years, surveillance and raids intensified feelings of fear and widespread terror 'marked' the city. Many of us who were witness to the horror remember vividly when paramilitary gangs stormed through the streets and the army closed neighbourhoods to create 'free zones' for their counterinsurgency operatives. To run these operations, the military relied on a certain kind of geographical knowledge, mapping their targets within the streets and buildings of the city.[6] The hunting for 'subversives' re-inscribed with new meaning many ordinary spaces where victims were seized or killed. We could trace a memory map of the repression, drawn with individual stories, about events grounded in particular sites such as clandestine prisons, buildings shattered by counterinsurgency operatives, classrooms, or even the city river, the Río de La Plata, into which prisoners were thrown during sinister 'death flights'.[7]

Therefore, I am particularly interested in the relation between physical space

Figure 8.1. *Desaparecido* Street Art at the site of the Club Atlético (clandestine prison/ torture centre) 2005. (*Photo*: Melissa Young, from the documentary film *Argentina – Hope in Hard Times*)

and events and in spaces where acts of memory are performed and memory is engraved. I thus propose to examine the symbolism of urban spaces as transformed and reconfigured by human rights activists, as spaces of changing and conflicting meaning and symbolism. Hence, this chapter focuses on 'extraordinary' events that give new meanings and re-signify the 'ordinary' spaces where they take place. The scenarios for the experiences I discuss here are urban spaces in Buenos Aires. The social actors are the *porteños*, its inhabitants[8] who, over the years, have demonstrated what Wainfeld (2004) argues is one characteristic of Argentine society: the fervent custom of taking to the streets noisily to put pressure on governments at the chant of '*Si éste no es el pueblo/ el pueblo dónde está*' (If we are not the people, then who are the people?).

Mothers of the Plaza de Mayo

When the Mothers of the Plaza de Mayo first got together, their purpose was to inform society that their children were vanishing. The task was to find ways to communicate their grievance in a way that would impact people and move them to action. Since their beginnings, the Mothers created a particular style aimed at shocking, disrupting and galvanizing a paralysed society. It was marked by their powerful presence in public spaces and by establishing Plaza de Mayo (*la Plaza*) and the streets as their territory. Theirs was an assertive conquering and remapping of urban territories through marches, mobilizations, and performances, aided by their use of symbols and symbolic spaces.[9]

For the Mothers, *la Plaza* became both a space and a language. 'In the square, facing power, let's get together and enter into *la Rosada* [the "Pink House", the name given to the government house that is painted in this colour] to demand our rights', proposed Azucena Villaflor de Vinzenci, their leader (disappeared in 1977) (*Madres de Plaza de Mayo*, No. 37). In choosing a site for their political struggle, the Mothers decided to take over a historical place: the square facing *la Rosada*, a location which goes back to the colonial days when the *Zócalo* or *Plaza Mayor* was the town's heart. Plaza de Mayo is a landmark of Argentina's political life and the site where significant events in the country's history have taken place, ranging from the first act leading to the independence from Spain in 1816 to the massive rallies of the Peronist movement between 1945 and 1955. By placing their resistance to repression in this veritable *lieu de mémoire*, the Mothers marked the space with an alternative historical meaning, as important as the previous layers imparted by the historic events that had contributed to its symbolic value, conflicting and unsettling its significance.

Furthermore, the Mothers defined their space in the arena that has traditionally been the site of the popular and the alternative. As Bakhtin (1984) notes in writing about the unofficial culture of the Middle Ages, its territory was the marketplace, where a special speech was heard, different to the language of the Church and institutions. Plaza de Mayo provided this informal democratic meeting space, a marked contrast from going to an institution and presenting an accusation 'with a desk in between' social actors. It permitted non-hierarchical links of communications and ongoing dialogue between the Mothers. The bonds among them are deeply rooted to that site. As they express: 'We felt all equals there. We felt free to communicate between ourselves. We [Mothers] made a communion between ourselves in the Plaza'.[10] In spite of the repression, the Mothers gathered in the Plaza but were constantly intimidated and attacked by soldiers and police. At the peak of the terror, their periodic marches became a constant reminder of the repression in this 'liberated' territory, even if only for some minutes every week. The identification of their struggle with this location meant that when they legally registered the group there was an obvious name for it: Mothers of the Plaza de Mayo.

Indeed, on Thursday afternoons, the Plaza is the Mothers' indisputable territory. Their weekly marches are complemented by an annual 24-hour resistance march. For the 1982 march (the last year of the dictatorship), the police impeded access to the Plaza. But the Mothers refused to give up recuperated territories and ventured into other urban spaces, holding the march in an adjacent avenue. It was the occasion when their supporters created the slogan proclaiming the Plaza as the Mothers' place: *La Plaza es de las Madres y no de los Cobardes* (The Plaza belongs to the Mothers and no to the Cowards) (*Madres de Plaza de Mayo*, No. 13).

Useful parallels can be drawn between the Mothers' style and the popular theatre performances that flourished in Latin America during the 1970s. This theatre became a political tool in the service of movements for social change. Its arena consisted of non-traditional spaces; streets and public sites became the stages where actors engaged passers-by as active participants in the performances (Boal, 1974). The Mothers literally staged demonstrations on any occasion where witnesses could observe them. As street theatre combines script with improvisation, the Mothers were able to respond creatively to external circumstances. If authorities asked one Mother for her ID, all the Mothers would hand in theirs; if one Mother was arrested, dozens of Mothers would declare themselves jailed. When Alfonsín, the first civilian president, cancelled a meeting with them, the Mothers staged a sit-in, turning the government house into a camp site where they stayed overnight.[11] While legislators were discussing impunity laws, the Mothers participated in a simulacrum of torture centres, installing replicas of cells where ex-prisoners wearing hoods were chained to each other. The streets facing the Congress thus became the stage for a performance recalling the inhuman conditions that thousands had suffered. The event brought the horror to the centre stage of legislative politics and the memory of these atrocities gave new meaning to this location.

How the Mothers managed to convey the presence of the *desaparecidos* in the streets is one of the most powerful visual images that they have created. For the first presidential elections after the dictatorship (1983), the Mothers resorted to a shocking expressive form. Large posters, each with a life-size silhouette symbolizing a *desaparecido*, with her/his name and date of disappearance, were posted on walls throughout the city. These thousands of silhouettes became witnesses to and participants in the electoral process, a silent but strong reminder to a population that was about to elect a government; for while people went out to vote, the *desaparecidos* had also taken to the streets. During the previous days, volunteers gathered in squares to stencil large silhouettes on old newspapers pasted together, so creating giant posters. In each figure, people wrote in the name of one disappeared person (*Madres de Plaza de Mayo*, No. 4). The ritual of producing the posters was an act of memory and this strategy turned public spaces into printing shops.

The Mothers have always been extremely creative in using graffiti to mark public territories. Buenos Aires' walls have traditionally been the 'popular newspapers' of those sectors without access to other media. The Mothers adopted this method deeply rooted in the national political culture by, for instance, symbolically re-appropriating historical buildings and installing in the Plaza de Mayo large replicas of *la Rosada* or the Cathedral, so people could write on them what they could not write on the actual buildings. The Mothers also set new parameters in pressing their agenda into official celebrations. On various occasions, they marked the spaces where military parades took place, painting street surfaces and sidewalks with large silhouettes or gigantic white scarves, the symbol of their struggle (*Madres de Plaza de Mayo*, Nos. 33, 36, and 44). By carrying out this 'decoration' at the last moment, the authorities had no time to erase it, resulting in images of soldiers marching over asphalt with painted accusations against them, such as boots interposed with white scarves. Ironically, the streets as spaces for honouring the military turned into scenarios to denounce its crimes.

Figure 8.2. The Mothers march over a painted white scarf in the Plaza de Mayo, 1998. (*Photo*: Lorena Riposati)

Figure 8.3. Preparing a stage at Plaza de Mayo for the events commemorating 1,000 *Jueves* (Thursdays) of marches by the Mothers of the Plaza de Mayo, 1996. (*Photo*: Susana Kaiser)

Hence, by appropriating, adapting, and developing new forms of alternative media the Mothers' activism transforms and symbolically gives new meaning to the urban landscape. Theirs is a struggle for territories that they feel belong to the people. They do not ask for permission, they act. Occupying the Plaza, having demonstrations, or tapping into public utility lines (e.g., for lights during marches) is justified under the premise: 'It's ours, let's take it'.[12] As I will show, the Mothers set precedents and established guidelines for the Argentine human rights movement. Their street presence, their taking over of the Plaza and other public spaces, their marking of recuperated or liberated spaces, and their visually compelling performances defined a style that other activists would adapt to their own circumstances.

HIJOS

The focus in this section is the late 1990s and the activities of HIJOS, an organization of daughters and sons of *desaparecidos*, political activists, and people forced into exile. The acronym stands for *Hijos por la Identidad y la Justicia contra el Olvido y el Silencio* (Daughters and Sons for Identity and Justice against Forgetting and Silence). The group was created in 1995, at a moment when Argentina was gradually and uneasily consolidating its transition to democracy. This organization soon achieved local notoriety for the specific characteristics of its activism, in particular for its *escraches*.

Figure 8.4. HIJOS at the Plaza de Mayo commemorating 1,000 *Jueves* of the Mothers, 1996. (*Photo*: Susana Kaiser)

Escraches are a new form of public protest. The word *escrachar* is an Argentine slang term meaning 'to uncover'.[13] For HIJOS, *escrachar* is 'to reveal, to make public the face of a person that wants to go unnoticed'.[14] *Escraches* are campaigns of public condemnation through publicized demonstrations, which are well covered by the media and aim to expose the identities of hundreds of torturers and assassins benefiting from impunity laws. In mobilizing public support and rallying communities to participate actively in their events, HIJOS rely on announcements placed in newspapers and flyers circulated in the city. Marchers invade the neighbourhoods where torturers live, and walk the streets carrying banners and chanting slogans such as '*Alerta, Alerta, Alerta los vecinos, que al lado de su casa está viviendo un asesino*' (Alert! Alert! Alert all neighbours, there's an assassin living next door to you!), or '*Como a los nazis les va a pasar, a donde vayan los iremos a buscar*' (Just like the Nazis it will happen to you, wherever you go we'll go after you). They inform the community about the atrocities committed by the former torturers, giving away flyers serving as fact sheets about the *escrachado* (person who is target of an *escrache*) – including photo, name, address, activities during the dictatorship, human rights abuses in which he is implicated, and current occupation. The demonstrations end in front of the torturers' homes with a brief 'ceremony' – a few speeches, street theatre performances, and music. Marchers then 'mark' the torturers' homes by spraying slogans on sidewalks and walls. Red paint – symbolizing blood – is also usually splattered on building walls.

The HIJOS are true political heirs of the Mothers and their *escraches* are more than traditional demonstrations. They represent a new and dynamic twist in the public challenge to impunity and political amnesia. The symbolically powerful tactics of bringing back the past into the contemporary public sphere compels society to face the consequences of its failure to administer justice and to define its position toward past human rights violations and ongoing campaigns for accountability.

HIJOS developed their strategies within a political and cultural environment of legalized impunity, characterized by a process that could be described as the 'normalization' or 'naturalization' of living with major human rights abusers. This unhealthy tolerance for criminality included the acceptance that torturers and assassins had the right to social spaces that are not behind bars. In 1998, when *escraches* reached their peak, hundreds of *represores* benefiting from amnesties were free to roam public places, invited as guests in television talk shows, had become 'democratic' politicians, and were even qualified as kind parents of children they kidnapped after torturing and/or disappearing their biological parents. However, several initiatives threatened their 'well-being'. In particular, developments in the globalization of justice had resulted in campaigns to extradite them to face trials in other countries.

In the general human rights climate of the time, there was a focused iconization of evil and only a few faces and names were well known – i.e., members of the military juntas. As these few became symbols of terror, hundreds of *represores* went unknown to the majority of the population. It thus became difficult to evaluate public reactions to their free use of public, social, and media spaces. Those who were recognized increasingly faced harassment and even physical assault. *Escraches*, thus, contested denial and ignorance by making people realize that those responsible for massive atrocities might be their kind neighbour or the father of their daughter's best friend. However, carefully conducted investigations to track down former *represores* preceded each *escrache*. This was a key 'intelligence' activity of HIJOS; for they wanted to uncover and shame as many *represores* as possible but avoid targeting innocent people.[15]

In aiming for public exposure and humiliation, the *escraches'* goal was to eliminate or limit the access to public and societal spaces that represores had gained. Their actions constituted a metaphorical repossession of the streets by freeing them from these criminals' presence. The strategy was a move to tear off the protective shield of anonymity behind which hundreds of torturers hid. In the absence of sentencing, HIJOS wanted to curtail the 'peaceful impunity' enjoyed by represores by ensuring that their neighbours knew their faces and their crimes as they met them daily.

Evaluating *Escraches*

The *escrache* was a very creative and powerful communication form, which, as I

Figure 8.5. *Juicio y Castigo* (Trial and Punishment). *Escrache* by HIJOS, 2001. (*Photo:* Lorena Riposati)

Figure 8.6. *[Si no hay justicia] Condena Social* ([Without Justice] Social Condemnation). *Escrache* by HIJOS, 2001. (*Photo*: Lorena Riposati)

discuss, was quickly adopted by other groups struggling for a variety of social justice causes. During 1998, I took part in two, and was able to observe the interactions between the marchers and neighbours, many of whom joined and celebrated the protests, and thanked the demonstrators for alerting them to the sort of neighbours they had.[16] A few neighbours told me how critical they were of the crimes committed by the *escrachados*. A woman even confided to me that she knew the whereabouts of a *represor* and was considering contacting HIJOS.[17] Others closed their windows, turned off the lights and stayed away but they did not defend the torturers.

Whatever the objectives of their organizers or my comments on their symbolic potency as communication strategies, the ultimate issue is how *escraches* impact the public at large. This is hard to measure because effects vary and may not be immediate. So far, we can only speculate about the consequences of augmenting the physical and metaphorical spaces in which *represores* were uncovered. But we can explore effects by assessing how people judge the strategy. Furthermore, attitudes toward *escraches* provide hints on how people experienced living in a culture of impunity and on their willingness to participate in ongoing human rights campaigns. So I researched how some young *porteños* 'read' the *escraches* and evaluated their effectiveness. I conducted interviews with young people who were either born during the dictatorship or after the commencement of civilian rule, and whose knowledge of that historical period was based on the stories that they were told.[18]

Although most interviewees rejected impunity and expressed sympathy for HIJOS, *escraches* were very controversial. Some believed that the dictatorship's crimes did not justify harassing these *represores*, even when there was awareness of the cruelty and sadism of these crimes and the explicit connection of these individuals to them – e.g., torture of disabled people, children witnessing the torture of their parents or being tortured themselves to make their parents 'speak', or special torments for pregnant women. There was a generalized condemnation for the *represores* but, overall, there was inconsistency between the anger levelled against the torturers versus the level of tolerance for them. The interviews revealed the many issues at play when evaluating *escraches*: whose interests they serve, their objectives, impact on the torturers' families and neighbours, the style of the demonstrators (often perceived as overtly aggressive), concerns over the damage to private property, and the relationship between demonstrators and the wider public.

Assessments of the Strategy

One pattern that emerged was interviewees' attribution of purely personal motives to the demonstrators, or the perception that *escraches* were motivated by the HIJOS's pain and anger. From this perspective, the strategy was frequently perceived as a 'healing therapy', with the various implications that personalizing the issue entails. Other comments, also focused on individual motivations, showed how the struggle for accountability is often seen as the victims' job. Some participants, who saw the relevance of the strategy and linked *escraches* with their role as prods to social memory, minimized society's responsibility or how society might support HIJOS's efforts. For instance, 'It has to do with memory, so people know. Really it's the least [HIJOS] can do, what's available to them' (Silvia). However, other remarks, highlighting the strictly personal motivations of HIJOS, indicated that their actions were also a desperate appeal for society's support, therefore noting human rights activists' isolation and frustration. 'It's [HIJOS's] way of doing justice, of writing what they feel, the pain that remains in them. Because they're struggling to do something and no one pays attention to them' (Elena).

There were, however, positive assessments of the strategy. Several participants praised its creativity and political effectiveness. Many interviewees gave concrete reasons for their relevance and were knowledgeable about the *escraches'* goals.

They highlighted the critical importance of uncovering *represores*. For once the community recognize the *represores* they become restricted in the spaces where they can circulate without being harassed; they may eventually find that they need to leave their neighbourhoods altogether. So although *represores* may not be incarcerated, they become, to a certain extent, prisoners in their own homes. This drastic curtailment of their access to social spaces in the urban landscape becomes a punitive measure, a *de facto* sentence for officially acquitted individuals. As I was told: 'Now the efforts are to try to exclude them from society. This is the aim of the *escraches*. To know that if you leave your house you have to watch out because a stone might hit your head' (Diego). A young woman who talked of the streets as a space to enforce accountability, probably the only recourse at the time, further reinforced this idea: 'It's street justice. I believe that according to the law they won't ever be jailed, so for as long as they live, let them live like this. Those against whom they do *escraches* would have to move' (Lucía).

Debate over the Impact on Torturers' Families and Neighbours

There were debates over *escraches*' impact on torturers' families and neighbours. Some participants indicated that the personal motivations of HIJOS conflicted with the rights of the torturers and those around them. In spite of understanding HIJOS's actions, participants expressed concern at what they perceived to be unacceptable levels of hostility. As one participant said: 'The guy who was a torturer obviously is a vicious person. But [the demonstration] is very aggressive. That person has children and neighbours, and it seems to me very harsh that they'd go and write: "Your neighbour is a torturer"' (Analía). While their families should not be held accountable for these crimes, torturers' use of their families as protective shields raises the issue of how people prioritize their concerns, weighing between the rights of a particular wife and child or the debt that the criminals owe to society.

Some interviewees, on the other hand, considered that the *represores* must be uncovered not only for society but that their families should also know about their crimes. There were also divisions around how interviewees saw torturers' families – either as accomplices or as totally ignorant. But there seemed to be a general belief that it was quite difficult for families, especially wives, to have ignored their husbands' activities. The reasoning was that they must have known and condoned what they did. As one participant put it: 'For me, they are all the same shit. It isn't true that [his wife/children] didn't know. It's like saying that the Germans were not aware of what the Nazis were doing' (Eugenia).

The impact on torturers' neighbours was another topic that generated disagreement, including how neighbours should react – i.e., active support, indifference, or rejection. Neighbours' attitudes towards these demonstrations should not be considered in isolation but rather as being representative of attitudes in society at large with regard to torturers' use of social spaces. In praising *escraches*, some highlighted that HIJOS warned neighbours of the possibility that they may be living next to assassins. Participants were aware that neighbours had joined demonstrators, proof that the *escraches*' strategy accomplished its goals of 'marking' and humiliating the *represores*.

However, some interviewees questioned demonstrations that affected others who were not targets of the protest. Protecting public and private property was a recurrent argument, mainly related to the spraying of graffiti on sidewalks and multi-family buildings. One of the organizers' goals was that neighbours would isolate the *represores* and force them to move; and the staining of walls had a role to play here. As a member of HIJOS told me: 'We hope that those in the building who don't want to keep paying for a paint job would tell the torturer that he should move'. Most interviewees, however, seemed to disagree with this particular aspect of the strategy.

Participants were often critical of damaging property shared by others. They compared aggression against private property with the dictatorship's crimes: 'I don't agree with what was done, it's wrong, obviously. But nor is what [HIJOS] are doing right. Besides, they write on the walls' (Liliana). 'I saw once on the news that someone's neighbours were complaining, "If it were his house and his sidewalk and [HIJOS] do that it's fine but not in a building where I pay the same maintenance fees"' (Eugenia). However, some participants noted the absurdity of comparing torture and killings with damaging walls and equating the right to private property with the right to life. Walls of a stained building can always be repainted, but nothing could bring back a disappeared parent.

Many of participants' comments suggested that the media played an important role in portraying how the community responded to *escraches*. Television broadcast of neighbours' reactions to the demonstrations apparently indicated that many people were more bothered by the turmoil than by having a torturer living next door, and tended to believe that it was neither their responsibility that torturers lived there, nor their duty to isolate them. References to neighbours' indignation at the disruption caused by demonstrators' screaming and banging drums, suggested that many were willing to tolerate the torturers' presence and would not force them to move out. As one participant explained, 'Neighbours are not to blame for what Videla did. They cannot throw him out of the neighbourhood. "It's OK, it's fair" [to condemn/shame the torturers] is what people said on television. "It's fair but it's not my fault that the guy lives here"' (Laura).

It is hard to assess the networks' criteria in their singling out of neighbours for interview, which comments to include, and how to edit the entire account.

Producing news is a multi-step process and there are editorial decisions for selecting comments 'in favour', 'against', or 'different views' to provide an 'objective' coverage. For I also heard of neighbours in favour of *escraches*: 'I saw on television that one woman who lives in the building went downstairs and joined the demonstrators because she did not know she had a dictator living next door' (Andrea).

I also made sure to ask how participants would react to an *escrache* aimed against one of their own neighbours. By personalizing the situation, I aimed to assess their attitudes towards current human rights campaigns in which they could take part. Some said that they would join the demonstrators and would not be bothered by the commotion or the spray-painting of walls. But I also heard comments revealing how a legacy of fear was affecting participation in political demonstrations. Although the years of terror were over, several interviewees referenced fear when speculating on what they would do.

This fear was founded on their accurate perception that *represores* still had the power to repress those who protest against them. Paradoxically, the criminals under siege requested, and received, police protection. Thus, some *escraches* were violently suppressed. For example, on the eve of the *escrache* at the house of an infamous torturer, he issued an open letter to his neighbours saying that the police would protect them from the demonstrators – implying that those who dared to remind society of his crimes would be repressed, which is exactly what happened.[19]

HIJOS's Ongoing Escraches

But repression did not intimidate the HIJOS. They have continued with their *escraches*, targeting not only *represores* but also their accomplices, in particular those who hold economic power. Since the dictatorship introduced the economic policies implemented in the 1990s, they have organized *escraches* connecting the past political repression and the present economic situation. These included *escraches* to whom they label *genocidas económicos* (economic genocidal agents) who implement policies that result in 'hunger, unemployment, economic "adjustment", wealth concentration, auctioning of state enterprises, and destruction of the domestic market'.[20] Thus, even when the main goal of HIJOS is the uncovering of *represores*, they have made a point of also shaming members of those sectors that condoned, collaborated, and benefited from the repression.

For example, there was the *escrache* at the National Museum of Fine Arts (July 2000) targeting Nelly Arrieta de Blaquier, from the institution's board. Her family, members of the country's ruling oligarchy, owns *Ingenio Ledesma* in the province of Jujuy. In that town, in a 'blackout' during a tragic night in 1976, dozens of workers and activists were kidnapped and an estimated forty people remain missing. Interestingly, the establishment newspaper *La Nación* wrote an editorial comparing HIJOS's campaigns to those of the Nazis marking the houses of the Jews. Also, for an anniversary of the military coup (March 2000), HIJOS organized *escraches* to Monsignor Aramburu, recalling the Catholic Church hierarchy's complicity with the repression, and to Roberto Alemann, an economist with the military regime responsible for the 'economic dictatorship'. Other *escraches* have targeted institutions that are headquarters of top figures from the financial establishment, many of them with proven links to the dictatorship and beneficiaries of its economic policies.

Furthermore, HIJOS have developed follow-up 'memory activities', so the momentum gained with the *escrache* does not fizzle out as a one-time affair. This way of securing a constant presence in a neighbourhood reaffirms the boundaries of these reconquered territories. One tactic is the 'mobile *escrache*', in which activists revisit the residences of several *represores* who have been already *escrachados*. Demonstrators drive around town on bikes, motorbikes, or in cars or chartered buses, and stop at different houses within a two or three hour span. During these brief appearances protesters cause commotion but restrain from spray-painting walls and delivering long speeches or performances. Another activity is to return to the neighbourhood in the days following an *escrache* to strengthen networks with the community. This 'post-*escrache* day' takes place in a public space and is coordinated by the '*Escrache* Table', a group formed by

Figure 8.7. Ledesma: We won't forget '76. HIJOS painting graffiti, 2000. (*Photo*: Lorena Riposati)

different organizations such as students or artists. HIJOS show the photos taken during the *escrache*, broadcast from site, and organize video screenings and/or live music performances. The *escrache* thus becomes an ongoing event, with active participation of the community, which broadens the territory of protest. It is no longer limited to the front of the *represor*'s residence and the route to arrive there, but includes parks or squares as new spaces to stage discussions of the past repression.

There are also international *escraches* that HIJOS's groups in other countries (many members grew up in exile) have organized outside Argentina. One *escrache* targeted Cavallo, a torturer at the School of Mechanics of the Navy (ESMA), an atrocious centre for torture and extermination. He was apprehended in Mexico (August 2000) and eventually extradited to Spain at the request of Judge Garzón. HIJOS Roma organized an *escrache* (January 2002) denouncing the meeting of Argentine Foreign Relations Minister Carlos Ruckauf and Italian Prime Minister Silvio Berlusconi as an encounter among 'mafia members'.

Overall, while the prospects for accountability were extremely limited, Argentineans shared societal spaces with hundreds of *represores,* who were, until the *escraches*, enjoying the impunity granted by amnesties. *Escraches* disrupted the process of this 'normalization' of living with major criminals. They became *lieux de mémoire* of the dictatorship, reminding society of the existence of 30,000 *desaparecidos* through the visible presence of their children and supporters. Until the nullification of the impunity laws in 2003, and the consequent revitalized expectations for justice, HIJOS should be credited with limiting the *represores'* social and spatial freedom. Many *represores* would not have been reluctant to leave the country for fear of being apprehended by Interpol (International Criminal Police Organization) – or leave their homes to be insulted and beaten by passers by – if it were not for the unyielding efforts of HIJOS. *Escraches* trapped torturers and assassins by building metaphorical – and mobile – jails in neighbourhoods throughout Argentina, so that *represores* recognized by a particular community were, in the hopes of the HIJOS, isolated within that environment. They also showed that in these times of 'cyber encounters' it is still effective to take to the streets.

Before moving to a more recent historical period, I should note that, in 1998, the Mothers of the Plaza de Mayo organized their own *escraches*. I accompanied them on two occasions. One was against the priest Graselli, a former navy chaplain accused of being an informer who had become the confessor at a private Catholic school for girls; the other targeted President Menem when the General Secretary of the United Nations visited Argentina. Thus, there has been an intergenerational exchange of techniques among relatives of the *desaparecidos*. While the Mothers influenced the format and strategies of the HIJOS, they also learned from the campaigns HIJOS developed.

The Twenty-First Century: The New *Escraches*, *Piquetes* and *Cacerolazos*

At the turn of the twenty-first century, the streets of Buenos Aires became the scene of massive popular demonstrations triggered by a new political crisis rather than by the settling of accounts with the past. *Porteños* made international headlines with their *cacerolazos* by taking to the streets *en masse* armed with saucepans which they loudly banged calling for the ousting of all those in office, in all branches of government: '*¡Que se vayan todos y que no quede ni uno solo!*' (Throw them all out!). Such was the anger and frustration with the political class. During the turmoil of 19 and 20 December 2001, demonstrators were eventually brutally repressed, but President de la Rúa was forced to resign.

Those were times characterized by an array of protests against the handling of several unfolding crises. Most of these struggles, however, were for basic human rights, including civil, political, economic, social, and cultural rights. People organized to demand jobs, food, housing, education, and to protest against the authorities' moratorium on bank withdrawals of their savings. Various actors from many social classes adopted, developed, and implemented new strategies to demand democratization through major restructuring of political institutions and

Figure 8.8. Street Art demonstration demanding the resignation of the Supreme Court, February 2002. (*Photo*: Mark Dworkin, from the documentary film *Argentina – Hope in Hard Times*)

the economy.[21] Demonstrators borrowed human rights activists' techniques for other causes not directly related to the dictatorship.

The *caceleros* were accompanied by the *piqueteros*, who initially represented the coming together of a movement of unemployed people laid off in the mid 1990s in oil-producing centres of the south and north-west of the country. But the events of 2001 generated a political environment in which the *piqueteros* were joined by different organizations and activists.[22] In Buenos Aires, the blocking was of streets and points of access to the city. Demonstrators chanted: '*Piquete y cacerola la lucha es una sola*' (Blocking and saucepan: there's only one struggle). This slogan conveys well the concept of a unified coalition joining forces to defeat the enemy, whether it be a corrupt government or unemployment and hunger.[23] *Cacerolazos* and *piquetes*, strategies involving the temporary ownership of territories with the purpose of causing disruption in everyday life and upsetting the *status quo*, became another means to transform urban spaces. Artists and musicians also participated in these popular mobilizations.[24] The performative angle of these massive demonstrations is another inheritance from activists such as the Mothers and the HIJOS.

The movement of *ahorristas* (holders of frozen saving deposits, mostly from the middle classes) adopted various strategies for occupying public spaces. There were, for instance, people who, unable to go on vacation, brought their chairs, umbrellas, and coolers and sat outside the bank, as the alternative to picnics at the beach. While these protests were localized in the financial district of Buenos Aires, the angry *ahorristas* also borrowed techniques from the *piqueteros* such as blocking streets in other areas of the city and expanding their territory of disruption.[25]

I should also mention the Neighbourhood or Popular Assemblies, which are meeting spaces for organizing that brought together large sectors of the community. These forums were key for analysing the crisis, deliberating about options, and making political decisions. Citizens discussed a broad variety of issues ranging from Argentina's foreign debt to unemployment, including the media, or the high fees of privatized public services (Kohan 2002, pp. 99–104). During those times (2002), the gatherings of the massive inter-neighbourhood association in Parque Centenario, located at the centre of the city, turned this large park into the 'legislative chamber' for a truly democratic experiment in grassroots democracy.

The New *Escraches*

HIJOS developed their *escraches* as a direct response to the impunity for the dictatorship's crimes. However, within this context of massive mobilizations of *caceleros, piqueteros, ahorristas* and *asambleístas*, there was also a new breed of *escraches*, which, inspired by and modelled after HIJOS's original strategy, was adopted as a means of denunciation and public condemnation by other organizations. This is a triumph for HIJOS and illustrates how society understands and approves of this tactic, and how human rights activists learn from each other.

The new *escraches* are significantly different. They are not directed against the *represores* but politicians, businessmen, or anyone in a position of power considered responsible for a current political or economic crisis. The target can be a former minister, someone responsible for layoffs, the owner of a supermarket chain that has drastically increased prices, or McDonald's restaurants, symbols of global corporations and their perceived negative impact on the local economy. They are not necessarily aimed at uncovering these people, who may already be well known, but rather to shame them. They thus protest against hardships in relation to a contemporary situation, usually without specific and/or apparent links with the dictatorship. Often, these demonstrations are not planned in advance. Rather they are spontaneous uproars of anger and frustration erupting when someone believed to be responsible for hardship is recognized in a public place. The fact that some spotted persons turn out to be *represores* is a testament to HIJOS's success in uncovering those benefiting from the impunity laws.

Some of the new *escraches* are related to the events of the dictatorship though. Malvinas veterans (from the 1982 war against Great Britain over the Malvinas/Falkland Islands) have 'commemorated' 'Malvinas Day' with an *escrache* to Galtieri, the dictator who led Argentina into the war. Others condemn official economic policies by targeting those responsible for them, as is the case of *escraches* to international visitors such as delegations of the International Monetary Fund who are *escrachados* at their arrival to the airport or at their hotels. There have been *escraches* to an employer for sexual harassment, to the director of a hospital who refused to perform an abortion on a physically and mentally disabled teenager who had been raped, or to corporations polluting the environment (Marabotto, 1999).

There have also been *escraches* to the media. For taking to the streets is intrinsically an effort to disseminate accurate information and counter the distorted versions of events offered by the establishment's media. When those making history are shunned by the media, they react against the apparatus that ignores their role as key protagonists. As Riposati (2003) explains, these escraches started after the events of December 2001 as a result of the massive popular participation in them. If the *piqueteros* staged a protest in which they blocked ten highways, television stations would report only on one while a major newspaper would not even mention the event. Faced with such experiences, dissatisfied neighbours from popular assemblies developed the idea of *escraches* to the media to denounce this biased coverage.[26]

The slogans painted on Buenos Aires' walls and the chants during these *escraches* summarize the accusations against the media, as illustrated by the blunt and explicit: '*Nos mean y los medios dicen que llueve*' (They urinate on us and the media say that it's raining).[27] This disparity between the media and the people in the streets, and the awareness of the ownership of well-defined territories, was expressed in the graffiti: '*La prensa es del capital, las calles son nuestras*' (The capital owns the press; we own the streets) (Corti, 2003). The perception that the media ignore and distort what happens in the streets prompted calls for being physically present at the locations of the events. One year after the repression of December 2001, Argentina Arde, a counter-information collective created after those events, covered the city with posters reading: '*El 19 y 20 apagá la tele y salí a la calle*' (On the 19th and 20th, turn off your TV and take to the streets).

HIJOS are aware that the protagonists of the new and often spontaneous *escraches* were inspired by and learned from them. However, they identify key differences. HIJOS's strategy continues to be consistent with their original goals: to uncover the *represores* from the times of the dictatorship and that these criminals be prosecuted. Their *escraches* become 'an alternative recourse to justice resulting from a consciousness raising process'. But they see the new *escraches* as a 'multi-directional expression of impotence' and a 'disorderly way of denouncing'. Furthermore, HIJOS argue that when the goal 'is not to achieve "popular justice", it risks fading into political emptiness'.[28] Thus, taking to the streets as a means to achieve 'popular justice' or 'street justice' is a distinctive characteristic of their use of public spaces.

There are many open questions regarding *escraches*' future. Will the new *escraches* remain spontaneous denounces? Or will they develop follow-up activities leading to an organized pursuit of justice, aimed at the prosecution and conviction of all those who abuse civil, political, economic, cultural and/or social human rights? What will be the links between physical and cyber spaces? Will the new media help to trace the geography of impunity? So far, the media have covered *escraches* widely and the Internet has been useful in coordinating and announcing public demonstrations. How will new technological developments further facilitate activists' work? Will it be possible to access a 'Yahoo.criminal' or a 'Yahoo.torturer' website and enter a zip code to locate all classes of criminals such as assassins, torturers, and corrupt politicians or businessmen who negotiated loans irrespective of their high social costs, or those who embezzled public monies and fled?

Epilogue

As I write down these thoughts, the urban spaces of Buenos Aires continue to be re-signified by ongoing protests of thousands of demonstrators taking to the streets. Therefore, these concluding remarks are comments about history in the making.

There are substantial differences, but also similarities between the Mothers of the Plaza de Mayo, HIJOS, and other demonstrators who take to the streets. These differences run across the organizers of the new *escraches* to the *piqueteros*, and including the *caceroleros, asembleístas,* or *ahorristas*. Human rights struggles are a common thread linking these movements, which also share a legacy of exclusion from the system and the official institutions. The Mothers took to the streets when authorities denied that the horror was happening. They were the main visible resistance to the repression by taking the issue of disappearances to the public sphere. HIJOS's *escraches* were developed to confront civilian governments' refusal to deliver justice. The strategy played a key role in challenging impunity and political amnesia in post-dictatorial Argentina. Most of the current demonstrators suffer from a lack of representation; they are mostly jobless and therefore cannot depend on unions to press for their rights. Their screams are loud and clear, but few in society are listening or responding to them.

Although the political environment is constantly changing, battles for human rights continue to take place in public spaces. A strong street presence is the quintessential characteristic of Argentine human rights activism. We talk of street demands, streets as spaces for deliberation and building democratic consensus, streets as arenas to denounce the injustices of the system and administer popular justice, streets as integral to dominated, intangible, and ephemeral *lieux de mémoire.*

Both the Mothers and HIJOS have defined a new style of demonstration, as part of a project of reshaping the uses of the public places of the city. The Mothers set the precedent for the 'ownership' of public spaces, in particular the Plaza. HIJOS's struggle for territories, for liberating the streets of *represores* now extends to freeing them from corrupt politicians and/or businessmen. Over the years, activists have been learning from different struggles and techniques, borrowing, co-opting, adapting, and innovating. The Mothers organize their own *escraches*, the HIJOS follow the path of the Mothers and adopt popular theatre, the frustrated *ahorristas* become *piqueteros*, and the *piqueteros* demonstrate with musicians supporting them. Thus, we can identify common traits across styles, extending to the aesthetic dimensions and performative connotations of these demonstrations. There is often a festive tone, highlighted by the presence of musical groups, actors staging street skits, or the screening of videos. This is made all the more crucial since human rights activists know clearly that in a bustling metropolis such as Buenos Aires they have to compete fervently for the public's attention.

It is difficult to measure how effective 'taking to the streets' has been in transforming social reality in Buenos Aires, and in Argentina at large, since the times of the dictatorship. However, when discussing human rights activism, the Mothers and HIJOS emerge as mandatory symbols of the resistance to impunity, as the embodiment of the struggle of memory against forgetting. Each gain in the pursuit of truth and justice is, to a great extent, the result of the relentless efforts by activists who skilfully combined investigations and legal recourse with their powerful street presence, a style that shaped their communication strategies throughout evolving political contexts. We may argue that if these activists had played according to the rules of the system they were questioning they would never have produced the necessary disruption to be noticed. It is hard to imagine that they would have had a similar impact on society if they had limited themselves to meeting in offices and issuing press releases. They would probably have become invisible to the public at large since the mainstream media would have ignored them.

Undeniably, their messages have influenced policy and action, placing and keeping the issue of human rights in the public sphere and on the politicians' agenda. But we can only speculate about their long-term impact. For these activists signify the popular conscience and memory, which are ethical and moral issues that have to be appreciated qualitatively, meaning that only a handful of protesters at a march can have a disproportionate impact on society as a whole.

Things have changed in Argentina in the last two decades, and there are significant differences between the dictatorship and civilian rule. But society is encouraged to hope for the future and shelve the past, to assume that along with the obsolescence of past terminologies – e.g., revolution – we need to dispense with ideologies, meaning that struggles for a more just social order have become 'old history'. Activists' loud street chants keep challenging the discouragement of criticism and the promotion of limitations, be it in justice, in what can be said, in what can be achieved, and even in what can be dreamt.

It is late afternoon in Buenos Aires and it will soon be the time for the legions of *cartoneros* (those collecting paper and cardboard) to start their nightly ritual of digging into the rubbish and gathering recyclable junk. Entire families left out of the system, reminding society of their poverty, turn the streets into their workspace, a large recycling plant. They hope to survive by collecting, classifying, and selling what others discard. It is hard to think of another situation that summarizes and symbolizes so well the cycle of exclusion. The *piqueteros* are protesting in the square facing the National Congress. The building housing the University of the Mothers of the Plaza de Mayo, one of the latest ventures of their creative activism, is part of the background scenario of this demonstration.[29]

Two decades after the commencement of civilian rule, I cannot avoid comparing initiatives that speak of exclusion and a generalized profound disillusionment with the political class and change that can be achieved through electoral politics. The

Mothers founded their university but zealously guard their 'liberated territory' and keep on marching every single Thursday. Unemployed workers, a heterogeneous group comprised of women and men of all ages, including their young children, roam the streets of Buenos Aires to expose their condition and struggle for their rights. Those activists seem convinced that the most relevant political spaces, maybe the only ones where they can articulate and debate their situation, are the streets, the logical and appropriate setting to exercise their political life. The streets of Buenos Aires are where it is all unfolding. Faced with the condition of being abandoned by the system, activists respond by being out on the streets, *¡Todos a la calle!* And I am drawn to conclude that human rights activists will continue to transform ordinary spaces with their extraordinary events.

Notes

1. Most of the section on HIJOS was published in Kaiser (2002).
2. 'Full Stop' Law No. 23493 (12/23/86) established deadlines for prosecutions. 'Due Obedience' Law No. 23521(6/4/87) exempted torturers and assassins from punishment by virtue of having followed orders. President Menem's Decrees of Pardon Nos. 2741-43 (12/30/90) pardoned the few *represores* still in prison.
3. Unsolved crimes include the assassination of photojournalist Jose Luis Cabezas, the bombings of the Embassy of Israel and the AMIA (association of Jewish organizations).
4. Comparable uses of public spaces by activists can be identified in cities and towns across the country.
5. See, among others: Downing (2001); Mattelart and Siegelaub (1983); Rodriguez (2001); OURMedia/NUESTROSMedios, a network of academics, practitioners, and activists supporting alternative media (http://www.ourmedianet.org/).
6. Harvey (2000, pp. 552–553) notes that 'geographical knowledges have often flourished in subterranean environments', namely in agencies who serve governments such as the CIA. He argues that this geography uses understandings of human histories within geographical frameworks 'to manipulate and deceive' and that it 'demonizes spaces and places for political purposes'. This misuse of knowledge is justified in the name of national security.
7. One project of *Memoria Abierta* (Open Memory) (www.memoriaabierta.org.ar), a network of human rights organizations, is an inventory of the memory sites of state terrorism.
8. This name refers to the port city of Buenos Aires.
9. For an analysis of the communication strategies developed by the Mothers see Kaiser (1993).
10. Hebe de Bonafini. Interview with author.
11. Bonafini. Interview with author.
12. Bonafini. Interview with author.
13. According to José Gobello's *Nuevo Diccionario del Lunfardo, 'escrache'* is a noun that Italian immigrants used, usually in a derogatory way, when referring to a photo or image of a person (Marabotto, 1999). *Lunfardo* is a colloquial language popularized by Buenos Aires working classes.
14. HIJOS website http://www.hijos.org/espanol/index.html.
15. Before the *escrache* against torturer Peyón (July 1998), an anonymous call warned HIJOS that Rolón, another torturer lived in the same building. HIJOS told me that they would not target Rolón without verifying the crimes in which he was implicated and that he indeed lived there. In February 1999, there was an *escrache* to Rolón at that location.

HIJOS took six months to conduct a thorough investigation, a sign of how responsible they were about publicly shaming only those guilty of atrocious crimes.

16. *Escraches* against General Riveros and torturer Peyón.
17. These conversations were during the *escrache* to Riveros.
18. I conducted individual and/or group interviews with 63 participants during May–August 1998, in the city of Buenos Aires. Most interviewees belonged to a wide spectrum of the middle classes, were not directly affected by the repression, and were not political activists. This research was for a study focusing on the post-dictatorship generation and identifying patterns of memory construction and transmission (Kaiser, 2000). It is also the topic of Kaiser (2005).
19. *Escrache* against Peyón, July 15, 1998. I witnessed this repression. Some Mothers of the Plaza de Mayo and several members of HIJOS were beaten.
20. Information about new *escraches* is mainly from *Indymedia Argentina* (www.indymedia.org) and it was accessed during March–May 2002. Because of the decentralized and independent posting of news, articles are often linked from other websites, posted more than once, or removed after a brief period, which makes precise citations difficult.
21. For the various movements generated by the crisis of 2001 see Melissa Young and Mark Dworkin's documentary film *Argentina: Hope in Hard Times*.
22. For an analysis of the development of the movement see Svampa (2003). See Kohan (2002) for a detailed chronology of the movements of *piqueteros* and *caceroleros*.
23. Gone are the days when everybody marched together. While *piquetes* have persisted, the *cacerolazos* movement has fizzled out (Corti, 2003). Wainfeld (2003) notes that the urban middle classes are furious with the *piqueteros* and severed alliances when their economic situation improved.
24. For musicians and performances in *piquetes* see Kohan (2002, pp. 129–136).
25. 'Unas 300 personas se movilizaron hoy'. Argentina Centro de Medios Independientes. http://argentina.linefeed.org/news/2002/05/26106.php. Accessed 11 May 2002.
26. I thank Lorena Riposati for updating me about *escraches* to the media. See also Riposati (2003).
27. For *escraches* to the media see videos produced by the collective *Cine Insurgente/Cine Piquetero*. Contact: cinepiquetero@datafull.com.
28. Indymedia www.indymedia.org. Accessed 10 March 2002).
29. The Mothers have created a university that grants diplomas and certificates for careers and seminars on a variety of disciplines and fields such as Human Rights, Social Psychology, Investigative Journalism, and Documentary Filmmaking. See http://www.madres.org/Universidad.htm.

Interviews

Hebe de Bonafini, president of the Asociación Madres de Plaza de Mayo. 6 May 1991.
Analía, 20 years, college student, communications major. 30 May 1998.
Andrea, 15 years old, high school student. 5 August 1998.
Diego, 21 years old, studying for teacher certification. 25 July 1998.
Elena, 15 years old, high school student. 24 July 1998.
Eugenia, 22 years old, student at medical school. 18 July 1998.
Laura, 19 years old, studying to be a history teacher. 25 July 1998.
Liliana, 15 years old, high school student at public school. 24 July 1998.
Lucía, 15 years old, high school student. 29 June 1998.
Silvia, 20 years old, studying for teacher certification. 23 July 1998.

References

Bakhtin, M. (1984) *Rabelais and his World.* Cambridge, MA: MIT Press.

Benhabib, S. (1992) Models of the public space: Hannah Arendt, the liberal tradition and J. Habermas, in Calhoun, C. (ed.) *Habermas and the Public Sphere.* Cambridge, MA: MIT Press.

Boal, A. (1974) *Teatro del Oprimido y otras poéticas políticas.* Buenos Aires: Ediciones de la Flor.

Corti, M. (2003) Ganar la calle: Arte y protesta política en la recuperación del espacio público. *Café de las Ciudades.* (http://www.cafedelasciudades.com.ar) Año 2, No. 2 (Febrero-Marzo)

Downing, J. (2001) *Radical Media: Rebellious Communication and Social Movements.* Thousand Oaks, CA: Sage.

Fentress, J. and Wickham, C. (1992) *Social Memory.* Oxford: Blackwell.

Galeano, E. (1989) *El Libro de los Abrazos.* Buenos Aires: Siglo XXI.

Habermas, J. (1979) The public sphere, in Mattelart, A. and Siegelaub, S. (eds.) *Communication and Class Struggle*, Vol. I. New York, NY: IG/IMMRC.

Halbwachs, M. (1992) *On Collective Memory.* Chicago, IL: University of Chicago Press.

Harvey, D. (2000) Cosmopolitanism and the banality of geographical evils. *Public Culture,* **12**(2), pp. 529–564.

González, J. (1999) Lectures at seminar on Research Strategies on Culture and Communication, University of Texas at Austin, July.

Kaiser, S. (1993) The 'Madwomen': Memory Mothers of the Plaza de Mayo. MA thesis. Department of Communications, Hunter College of the City University of New York.

Kaiser, S. (2000) *De Eso No Se Habla:* Transmission of Silences and Fragmented [Hi]stories in Young Argentineans' Memories of Terror. PhD dissertation. University of Texas at Austin.

Kaiser, S. (2002) *Escraches:* demonstrations, communication and political memory in post-dictatorial Argentina. *Media, Culture and Society,* **24**(4), pp. 499–516.

Kaiser, S. (2005) *Postmemories of Terror: A New Generation Copes with the Legacy of the 'Dirty War'.* New York: Palgrave Macmillan.

Kohan, A. (2002) ¡A las Calles! Una historia de los movimientos piqueteros y caceroleros de los '90 al 2002. Buenos Aires: Colihue.

Mattelart, A. and Siegelaub, S. (eds,) (1979, 1983) *Communication and Class Struggle*, Vols. I and II. New York: IG/IMMRC.

Marabotto, E. (1999) La exportación del escrache. *Clarín Digital,* http://old.clarin.com/suplementos/zona/1999/02/14/I-01401e.htm, 14 February.

Nora, P. (1989) Between memory and history: Les lieux de mémoire. *Representations,* **26**, pp. 7–24.

Nora, P. (1996) General introduction: Between memory and history, in Nora, P. (ed) *Realms of Memory: The Construction of the French Past,* Vol I. New York: Columbia University Press.

Rappaport, J. (1990) *The Politics of Memory. Native Historical Interpretation in the Colombian Andes.* New York, NY: Cambridge University Press.

Riposati, L. (2003) La Experiencia de Escraches a los Medios en Argentina. Paper presented at the conference OURMedia/Nuestros Medios III, 17–21 May Universidad del Norte, Barranquilla, Colombia.

Rodriguez, C. (2001) *Fissures in the Mediascape: an International Study of Citizens' Media.* Cresskill, NJ: Hampton Press.

Ryan, M. (1992) Gender and public access: Women's politics in 19th century America, in Calhoun, C. (ed.) *Habermas and the Public Sphere.* Cambridge, MA: MIT Press.

Svampa, M. (2003) El movimiento piquetero en Argentina: dimensiones de una nueva experiencia. Paper presented at the conference The Argentine Crisis. Indiana University, Bloomington, 29–30 September.

UNESCO - Intangible Cultural Heritage (http://portal.unesco.org/culture).

Wainfeld, M. (2003) Cuando los hermanos no son unidos. *Página 12 on line,* 30 November. (http://www.pagina12.com.ar/diario/elpais/1-28745-2003-11-30.html).

Wainfeld, M. (2004) La crisis energética y la demanda de seguridad jaquean al gobierno. *Página 12 on line 4* April. (http://www.pagina12.com.ar/diario/elpais/1-33695-2004-04-04.html).

Newspapers

Madres de Plaza de Mayo, publication of the Asociación Madres de Plaza de Mayo.
Página 12 online (www.pagina12.com.ar).

Acknowledgements

I thank Clara Irazábal and Macarena Gómez-Barris for their feedback on an earlier version of this chapter. I am grateful to Lorena Riposati, Melissa Young, and Mark Dworkin for allowing me to use their photos.

Chapter 9

Iconic Voids and Social Identity in a Polycentric City: Havana from the Nineteenth to the Twentieth Century

Roberto Segre

Throughout the history of Havana, the connections between public space and social life were manifest in a succession of relationships. During the colonial period, each of the various spaces of the existing urban polycentric structure was a stage for its share of the social functions of the time. The Cathedral Plaza, for instance, staged religious duties; the Plaza de Armas staged civic and military events; and San Francisco and the Plaza Vieja, provided commercial space. Each of these places was readily accessible to the working classes. This role for the popular spaces in fostering social integration remained intact until the eighteenth century. In the nineteenth century, however, the *criollos's*[1] struggle for Independence from Spanish colonial domination resulted in a territorial and ideological separation between national groups. The extraordinary events served as precursors to the wars of liberation whose drama unfolded in a somewhat removed area that would later be known as the *Parque Central*, where the armed revolutionaries clashed with their enemies outside the walls of the city.

During the era of the Republic, in the first half of the twentieth century, a modern Havana emerged, and care was taken in pursuing design projects that bestowed upon the city a scenic character. This was partially the work of a bourgeois ruling elite who adorned the city with architectural symbols of their power in the form of monumental public spaces in which they played out their social life. The 1950s were witness to the marked expansion of North American tourism, the blatant expression of tropical hedonism and the hegemonic influence of the automobile that promoted the suburban expansion. Except for the popular street carnivals, the use of public space was generally limited to socially oriented demonstrations against the successive dictatorships that characterized this period. With the advent of the Revolution in 1959, the streets, parks, and public squares hosted a populace whose enthusiastic popular participation in the political and cultural events that took place throughout these tumultuous four decades have changed the face of the city and country at large.

The socialist system, characterized by a top-down authoritarian political structure, leaned on popular support to legitimate decisions and radical initiatives. The multitudes attended the demonstrations, marches, and parades in some of the main iconic spaces of Havana: Plaza de la Revolución, the University Stairs, the Avenida de las Misiones, and the Malecón. In the second half of the twentieth century, Havana's urban history was defined by these significant political and ideological events that took place in the city's traditional public spaces. What the twenty-first century holds for the city after this momentous and tumultuous period remains to be seen.

Dynamics of Public Spaces

In Latin America, the Iberian conquest and colonization defined the city's public spaces. In some cases characteristics of pre-Columbian urban public space were maintained and in others hybrids were formed. According to Jorge Enrique Hardoy and Carmen Aranovich, more than 300 settlements were founded in the Spanish colonies over two centuries in the continent and the Caribbean (Hardoy and Aranovich, 1969). The design and planning of these settlements veered little from the norms that were institutionalized in the Laws of the Indies in 1573, during the reign of Felipe II.[2] The main city plans and the social functions that took place within them turned out to be very similar to the existing models which were originally codified in the Middle Ages. There was a *Plaza Mayor* in each, around which basic social functions took place: the town hall or municipal government, the church, commercial and financial institutions; and in port cities, the customs office.

It is through these structures that the Spanish colonial rule was symbolically represented and extended its reach to the peripheral indigenous nuclei (Rojas Mix, 1978). In isolated cases, new spaces were superimposed over pre-existing structures that were remnants of local civilizations, such as the *Templo Mayor* of the Aztecs in Tenochtitlan, which was refashioned as the Zócalo Plaza. In the majority of vice regal capitals – later transformed into the nascent republics' capitals – the space of the Plaza Mayor maintained considerable social, political, and cultural significance. Such were the Plazas Bolívar in Caracas and Bogotá; San Martín Plaza in Lima; the Plaza de Armas in Santiago, Chile, and Plaza de Mayo in Buenos Aires (Terán, 1989), among others.

Havana constituted a unique case with respect to the majority of Spanish colonial cities in the Americas. Founded within the interior of a bay by a small group of inhabitants – in 1538 it possessed only 70 households (Venegas Fornias, 2003) – the nascent Plaza Mayor lacked a defined perimeter. The main church and the precarious homes in its vicinity were situated near the edge of the bay. Even before the port became important to the navigation of Spanish ships on their return from the Spanish peninsula, pirates had begun to harass the city. In 1555 the city was sacked and plundered by the French privateer Jacques de Sores. Because of the need for protection, the Crown decided to construct the La Fuerza castle in front of the plaza in 1558. A space for military manoeuvres was left undeveloped, and a majority of the houses were razed. The church survived because of its marginal position with respect to the fort. After the futility of the castle as a means of defence became apparent, the residence of the governor was established in its second floor. The Plaza de Armas did not become part of a coherent cityscape and of social urban importance until the second half of the eighteenth century – through an initiative of the Marquis de la Torre (Venegas Fornias and Núñez Jiménez, l986) and with the construction of monumental palaces at Segundo Cabo

Figure 9.1. Aerial view of Plaza de Armas in the nineteenth century, drawn by Francisco Bedoya. (*Source*: Roberto Segre's archive)

in 1770 and Capitanes Generales in 1776. Its classic elegance, only appropriate to the status of the political and administrative headquarters in the city, was complemented by the already established connection with the main square through wide portals. These provided pedestrians protection from the blazing sun and torrential rains (Segre, 1996).

When the social function of the main public area in the city was curtailed by the building of the castle, the Town Council proposed, in 1559, to build a non-central plaza outside the compact area of the city. This plaza would serve as a location for the market and provide space for social and recreational activities. Hence, the *Plaza Nueva* (New Square), presently known as the *Plaza Vieja* (Old Square), was born. This first spatial and functional displacement initiated the development of a dynamic and polycentric Havana, a process that would accelerate in the twentieth century. Up until the eighteenth century, life in colonial society played itself out within the colony's walls. The public spaces inside the walls comprised the new Plaza de Armas, headquarters of the colonial government used by the members of Havana's high society for their evening strolls; Plaza de la Catedral, an urban space dedicated to the celebration of the affluent strata's religious ceremonies; Plaza Nueva with the market, surrounded by luxurious mansions with porticos; and the Plaza de San Francisco (San Francisco Square) which faced the harbour. Its layout was irregular and served as headquarters of the customs office and the town council and was also the site of popular carnival parties. Although a certain degree of separation between urban functions was maintained, a fluent and well-balanced relationship existed between the public and private spaces. These were favoured by a society that was racially heterogeneous and that favoured festive and business activities over those of a religious and administrative nature.[3]

Social and Ideological Antagonism in Colonial Havana

The independence movement that swept the Latin American continent and culminated in 1824 with the battle of Ayacucho, Peru had an important influence on Havana. The Spanish Crown had retracted and consolidated its presence to form a stronghold in Cuba and Puerto Rico in hopes of recovering its lost territories. Spanish businessmen and landowners emigrated to the capital of the Antilles and developed profitable businesses linked to the slave trade from Africa that was supported by the local government (Guimerá and Monge, 2000). At the beginning of the nineteenth century however, a *criollo* middle class had emerged. This class questioned slavery because of its inefficient productivity. Instead, they encouraged the use of steam engines in the sugar mills and demanded more political participation and commercial and cultural freedom. Prominent plantation owners and intellectuals such as Francisco de Arango y Parreño, José

Antonio Saco, José Agustín Caballero, Domingo del Monte, Tomás Romay, and the plantation owner Domingo Aldama were influenced in part by the writings of the Baron von Humboldt who wrote about the Cuban island (Portuondo, 1962; López Segrera, 1989). They joined in the criticism of the repressive politics practiced by agents of the Spanish government in the colony and radically carried out by Governor Miguel Tacón (1834–1838). The clash of interests was aggravated by the antagonism that had already been brewing between sugar and tobacco plantation owners. The latter identified more with the *criollos* and actively participated in the economy of Havana during the second part of the century (Ortiz, 1963).

These social and economic discrepancies became manifest in the urban spaces of the city. The city within the walls could not accommodate the growing population. Towards the end of the eighteenth century some settlements started to form outside the walls. Sunday strolls were taken at Alameda de Extramuros (a public walkway lined with trees outside the wall), previously known as Paseo de Isabel II and presently called Paseo del Prado. With the arrival of Tacón the ideological tensions between the *criollos* and the Spaniards assumed expression in urban development. During the brief period of his rule, his radical incisions into the city fabric, which outlined the future expansion of the city, earned him the reputation of an incipient Baron Haussmann (Chateloin, 1989; Segre, 2002). He equipped the city with the necessary social services – theatre, markets, hospitals, jails – and encouraged the expansion of Havana outside its walls. The open space of Campo de Marte, at the end of the Paseo de Isabel II linked to Tacón's new theatre and to Reina Street, and continued on to Paseo de Tacón. This was intended as the site for luxurious mansions, but was a failed project due to its rejection by the elite of the *criollos*, who later established themselves in the neighbourhood of Vedado. At the same time, summer residences were sprouting in distant Cerro, which was isolated from the bustle of the city and enjoyed better sanitary conditions. With time, these buildings were converted into permanent residences for the upper classes.

The ideological antagonisms became manifest in architectural and sculptural works in the public spaces. In 1828 Bishop Juan José Díaz de Espada y Landa sponsored the construction of a small temple in Plaza de Armas, meant to commemorate its founding and the first mass celebrated in the city. Its neoclassicism – the first for this style in Havana – resembled that of a similar building in Guernica, a Basque city traditionally rebellious towards the central authority in Madrid. Hence, a symbol of emerging anarchistic ideals circulating in Europe challenging the repressive rule of Fernando VII was erected in the hub of colonial power (Ortiz, 1943; Segre, 1996). Soon after, a rivalry arose between Governor Tacón and Claudio Martínez de Pinillos, Count of Villanueva, general superintendent of Hacienda and a figure sympathetic to *criollos*'s interests. In 1836, Martínez de Pinillos erected a marble fountain in front of Campo de

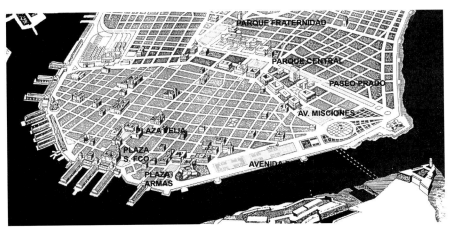

Figure 9.2. Perspective drawing of Havana, drawn by Francisco Bedoya. The city's public spaces are highlighted. (*Source*: Roberto Segre's archive)

Marte, a symbolically repressive location dedicated to military exercises. The Italian sculptor Giuseppe Gaggini was commissioned to design the fountain. It represented 'The Noble Havana', a slender figure of a beautiful Indian girl, later known as *Fuente de la India*, a metaphor of the autochthonous origins of the Cuban people (Roig de Leuchsenring, 1964; Gutiérrez Viñuales, 2004). He also placed another work by the same artist in Plaza de San Francisco, facing the harbour. The *Fuente de los Leones*, a representation of the umbilical cord with the motherland, alluded to the combatant spirit of the *criollos*.

Havana was polarized because of the domineering presence of the *criollos* in the area outside the walls. As a testament to this polarization, Domingo Aldama's monumental neoclassical residence, built in front of Campo de Marte, stood as a symbol of territorial claims. The Spaniards also occupied part of this area and were never shy to show disdain towards the subversive meetings held in El Louvre and Escauriza coffee shops and the theatrical productions in the Villanueva Theatre, which ridiculed the Spanish power.

The thriving social life which existed in the town squares in Havana until the eighteenth century was thus moved to the strip outside the walls (Venegas Fornais, 1990). This was energized by the presence of tobacco factories and workers' houses, which made up the neighbourhoods extending towards the west. During this time, local residents and tourists, most of them from the United States, frequented coffee shops such as the Paris, and the Alemán, patronized theatres such as the Villanueva, the Payret, and the Jané and stayed in such hotels as the Inglaterra, the Telégrafo, the Pasaje and the Plaza. Each of these structures surrounded the future Parque Central. Without a doubt, the Plaza de Armas was a symbol of the political and ideological contradictions existing in the colonial era.[4]

In a country that was plagued by a decades long war of liberation, and in a city that was divided along social lines, public space in Havana provided little sense of community. While the Central Park attracted the beggars, impoverished by the cruel policies of 'concentration'[5] of Captain General Valeriano Weyler (Pogolotti, 2002), the aristocratic Plaza de Armas was the stage for the first populist gatherings of Cubans. Twice emotional crowds gathered in the plaza. The first time was to provide a magnificent welcome to Generalísimo Máximo Gómez, the hero of the War of Independence, who was later forsaken for subordinating to the political interests of the United States. On the other occasion, crowds gathered for the funeral of General Calixto García, also a participant in the War of Independence and a negotiator for the incipient Republic which had faced the annexation attempts by its 'Northern Giant' (Bretos, 1996). Although during the first decades of the twentieth century it housed a strong concentration of banks, hotels, offices and public buildings, it was not this area of colonial Havana that would attain the greatest symbolic value in the social imagery.

During the republican period (1902–1958) Havana was not a city supporting massive popular activities in its public spaces. On the one hand, the city planning inherited from the colony, which basically created open spaces that framed the public buildings, did not provide an environment with strong iconic value that would be meaningful to dissimilar social groups. Parque de la Fraternidad (Fraternity Park) was one of the main centres where the Sixth Pan-American Conference (1928) was officially celebrated, an event attended by Calvin Coolidge, President of the United States. The hypothetical continental alliance was symbolized with the planting of a ceiba tree with soil from all Latin America countries. On the other hand, political vicissitudes created by a corrupt and repressive system of taxation and limited democratic participation were exacerbating social rifts. Although the first organized labour unions appeared in the 1920s, a proletariat presence during political rallies was limited. These were typically repressed by the ruling government, especially during the dictatorship of Gerardo Machado (1925–1933).

It was through an initiative of the state after World War II that the building of the monumental and classic elements of the 'spectacular city' (Álvarez-Tabío Albo, 2000) began in earnest. This was reflected in the public institutions that were copied from European and North American prototypes, such as the Presidential Palace, the National Capitol, and the University of Havana among others. However, at the beginning of the century, the US Army Corps of Engineers built Gulf Avenue (better known as the Malecón) which served as a barrier between the city and the sea. This became an area intensely used by the growing middle class but scarcely at all by the low-income population of that time.

The Malecón traversed popular neighbourhoods such as the centre of Havana and well-to-do neighbourhoods such as Vedado. It became attractive to those seeking cooler temperatures and sea breezes. At the same time, the elite classes began to leave the central areas along Paseo del Prado which they had occupied since the beginning of the century. Instead they moved towards the west and began to build their residences in the new neighbourhoods of Jardín del Vedado, Miramar and the Country Club. These were close to the coast and became classic dormitory suburbs.

The Central Park became the hub of the city whose periphery extended into complementary spaces. In one direction was Paseo del Prado and the Presidential Palace with Avenida de las Misiones (Missions Avenue). In an opposite direction was Parque de la Fraternidad. This park was identified as the starting point of the *criollo* anarchic struggle against repression and was reflected in a monument dedicated to José Martí. Although the incipient republican democracy was mainly symbolized by the dome of the National Capitol, the monumental presence of Centro Gallego and Centro Asturiano made clear that the political and social antagonisms had vanished.

This coincided with the financial interests of the Cuban, Spanish, and North American elites. The diversity of functions present in the Central Park area – masterfully described by Alejo Carpentier (1959, 1996) – symbolized the vitality and dynamism of the urban life. This was expressed by the mixture of social groups – represented by Walker Evans's compilation of photographs, published in 1933, in the books *Citizen in Downtown Havana* and *Havana Corner* (Evans, 1933, 2001). In August of that year that enthusiasm was put to action when an excited group gathered in the centre of the city, from the Presidential Palace to the Capitol to commemorate the end of Gerardo Machado's dictatorship (Le Riverend Brusone, 1966).

Cuba's capital – which in the 1930s reached a population of half a million – was characterized by its pedestrian life which resulted from the system of plazas that configured the compact colonial zone, carried on in Central Park and expanded along the covered sidewalks. This feature was incorporated by the French urban planner Jean Claude Nicolas Forestier, who was invited by Carlos Miguel de Céspedes, Secretary of Public Works in Machado's government, to design a new master plan of Havana (1925). When the streets were mapped out with diagonals, roundabouts, and avenues with generous parterres in the green areas downtown – from Fraternity Park to Avenida del Puerto (Port Avenue) – he designed the urban 'salon' of Paseo del Prado, protected from the tropical sun by luxuriant oleanders. The plan was sensitive to the ecological and climatic requirements of a tropical metropolis and reproduced images of boardwalks in Mediterranean cities such as Niza, Málaga, and Naples (Lejeune, 1996; Segre, 2002). In addition to the Malecón, Forestier designed the Avenida del Puerto with the intention of creating

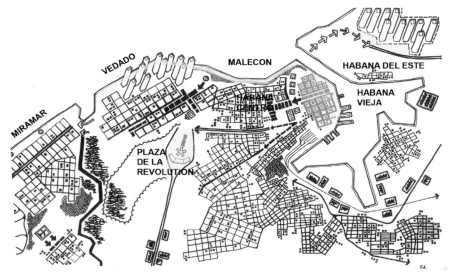

Figure 9.3. Plan of metropolitan Havana. Old Havana, the Malecón, the US Embassy, and the Civic Plaza (later Revolution Plaza) are highlighted.

an extensive public space, annexed to the compact colonial city. It was intended for the relaxation of the less affluent classes. An open amphitheatre was later built on that site. These urban enhancements made Havana all the more attractive to the growing US steamship industry that increasingly brought American holidaymakers to the city (Schwartz, 1997).

However, the elites did not look favourably on the mixing of the classes in the downtown area. The main tobacco factories, Partagas and La Corona, were located in front of the Capitol and Presidential Palace, whose expansion was limited by their proximity to the poor neighbourhoods. Thus, building a new political-administrative centre in a less developed area of the city was proposed. This was La Loma de los Catalanes. It was adjacent to Vedado and would allow the creation of the monumental Plaza Cívica (Civic Square) framed by the new governmental buildings and the expansion of future middle-class neighbourhoods away from the poorer sectors. Although Forestier's project did not come to fruition because of the overthrow of Machados's government, the site was recognized as the focal point of new metropolitan Havana.

Pleasure and Cruelty in Havana's Nights

At the end of World War II, Havana radically leapt towards modernity. On the one hand, social disparities increased. For instance, the dictatorship that started with Fulgencio Batista's *coup d'état* in 1952 made for a distressing grip of urban violence on city streets. On the other hand, an economic boom was engendered by the sale of sugar to the countries involved in the war and by the sizable American investments in the 1950s. Together, these helped to transform Havana into a resplendent, luxurious, and vibrant city, criss-crossed by wide avenues that framed buildings of superior architectural design. These originated from the efforts of a small, local, and refined middle class who endorsed the aesthetic innovations of the young professionals working in accordance with the precepts of the Modern Movement (Rodriguez, 1998, 1999). The automobile was another essential element in the transformation of the traditional character of Havana which previously had been defined by its compact structure. With the advent of freeways, the city expanded towards the open peripheral areas where residential neighbourhoods were located. In the 1950s, Cuba boasted the highest number of cars per person in Latin America. This transformation is made evident in the works of its literary interpreters. While in the 1930s Alejo Carpentier, local *flâneur*, wandered on foot to describe the Central Park and the 'city of columns', in the 1950s Guillermo Cabrera Infante did the same, driving at high speed in his luxurious Cadillac. This was possible because of the 'centrifugal' structure of the new suburban neighbourhoods (Cabrera Infante, 1987; Izquierdo, 2002). It is not by chance that the Malecón eventually became a first class speedway.[6]

A city planned to accommodate millions of American tourists did not favour the development of public spaces for the local residents. In the 1940s, during the government of Ramon Grau San Martín, numerous urban parks were built. These constituted open space for the speculative expansion of new neighbourhoods without any particular symbolic significance. The street life was divided into two main areas. The first comprised the traditional historic downtown area, which was integrated with the harbour area. This area was characterized by closely packed popular businesses that branched out into the arcades of the surrounding buildings and extended to the big department stores of Calle Galiano (Galiano Street). The second hub was in the incipient development of Vedado which, dominated by the renowned Hotel Nacional (built in the 1930s), boasted the 23rd Street (also known as La Rampa) with its hotels, night clubs, cinemas, and offices. At the same time, tourist and recreational activities were decentralized and moved to the luxurious and aristocratic coastal clubs, the remote cabarets – e.g., the memorable Tropicana – as well as to the adjacent beaches. Luxury yachts that set sail in Miami docked in Barlovento and Marianao, bringing their wealthy passengers to the tropical nightlife.

Hence, in the 1956 Main Havana Plan, designed by the New York-based planning and development office of Lester Wiener, José Luis Sert and Pual Schulz (recognized as the city planning spokesmen of the Modern Movement in Latin America) there were no provisions for public spaces or for public gatherings, putting emphasis on the road configuration geared for the use of the automobile (Rovira, 2000). In turn, the historic downtown area was eliminated (Segre, 1995) and transformed into a Hollywood style space for tourists. The Malecón was dotted with high-rise luxury hotels and apartments, totally losing its already scant public significance. Forestier's academic design of the Civic Square was also 'modernized'.[7] This fragmented its unitary configuration in an asymmetric articulation of public buildings and smaller plazas, although the presence of the José Martí Monument and the symmetrical Palace of Justice – both built by Batista's government – established an axial organization of the huge space. In this space were buildings of different styles, such as the National Theatre, Ministry of Public Works, Bus Terminal, National Library, and Ministry of Communications and Town Hall. These structures consequently stopped the square from ever achieving a coherent form, while continuing to be the biggest urban void of the capital. It was eventually used for political rallies organized by Fidel Castro's government. With the exception of the road system and the urban development in east Havana that were accomplished by the revolutionary government, albeit within a different social context, these were proposals which did not make it past the blue-print stage.

In the 1950s, Havana did not have a unique public space that integrated social, cultural, and political activities. Carnival, celebrated in Fraternity Park and the Malecón alongside Central Park drew the biggest crowds. On the anniversary of the birth of José Martí, his statue in Central Park would draw school students who would march through the city streets (Chacón Nardi, 2002). That was the site of the infamous incident in 1949, where a group of drunken American marines got on top of the statue and urinated on it. This precipitated a popular reaction that culminated in an angry protest headed by the young attorney Fidel Castro – in front of the United States embassy, located in Plaza de Armas.

The front steps of the University of Havana in Vedado were also a site that gained important symbolic meaning. Students there constituted the main segment of society active against the successive dictatorships. Before Batista permanently shut down the University in 1956, student demonstrations would proceed from the hill down the monumental front steps towards downtown. During the celebration of the centennial anniversary of the birth of José Martí in 1953, and in opposition to the official events organized by the government, 500 students with torches started their march on the front steps and lit the city on that dark night in January (Raffy, 2003). A few months later another multitudinous march accompanied the funeral of student Ruben Batista Rubio – murdered by the police during a street demonstration – from the hill to the Colón Cemetery. Finally, the architecture student and leader of the University Student Federation, José Antonio Echeverría, was murdered close to the front steps by the repressive forces on the day of the unsuccessful attempt to assassinate Batista at the Presidential Palace (1957). With all these extraordinary events the city oscillated between hedonism and protest. On the one hand this proceeded from the frivolous life of the local elites as well as the American tourists in the 'smart' areas of the city. And on the other hand it proceeded from the anguish caused by the poverty and wretchedness of the precarious suburbs and the ruthless street killings of daring young men willing to sacrifice themselves for the restoration of democracy. This was a tense reality that was radically changed in 1959 with the flight of Batista and the defeat of his army in Santa Clara by the guerrilla force commanded by Ché Guevara.

Community Presence in Urban Spaces

If during the span of more than four centuries Havana's public spaces were mere reflections of the functional and symbolic city planning and architecture of the day, in the last half of the twentieth century they were characterized by the dynamism of popular participation and the multiplicity of functions, events, celebrations, marches, parades, and festivities that animate place in the contemporary urban context. These social forces appropriated the symbolic spaces in the city. For instance, the capital became the stage for the political life of the country. The revolutionary forces mimicked the October Revolution in Russia, in which the masses constantly congregated in streets and plazas. The intention was to forge a socialist utopia which replaced individualism with group participation and inward-looking family life with the integration of the masses in the community. Hence, streets and monuments became the silent testimony to the heroic and fiery acts of the revolutionary process (Vidler, 2004). This silent city planning took on an anti-urban undertone when confronted with the migration from the countryside.

The social occupation of the urban space by the newly arrived resulted in the loss of that space's identity with its diverse cultural and architectural values. These should have been maintained and multiplied. Instead, an urban identity was suppressed with the 'ruralization of the city' – a consequence of the presence of a significant segment of the rural population in Havana. This population transferred rural customs and practices to the urban setting. This became commonplace with the new constructions built for the residents coming from the interior of the country – the nondescript blocks of homes – built in different neighbourhoods (Baroni Bassoni, 2003) that lacked a feeling of identity. This new construction accelerated the general physical deterioration, and these newcomers formed an urban sub-culture whose dire fortunes were aggravated by the economic crisis of the 1990s (Coyula, 1996).

In four decades little or nothing was built in the inherited urban grid, with the exception of some groups of houses, schools, orphanages, hospitals, in a tendency to create new symbolic sites for community activities. In this sense, Cuban socialism differed drastically from the Russian prototype. It shunned the construction of monuments symbolizing the new social and political structure and of government power. In Havana, with the disappearance of private enterprise, the state institutions occupied the already existing public buildings and many of the modern offices. In turn, there was always an attitude of disdain towards the inherited city by the political high command, as it was regarded as a representation of the vices and injustices present in the capitalist system. Therefore, the new educational, housing and recreational infrastructure was mainly placed in the urban periphery. These included the José Antonio Echeverría University, the W.I. Lenin Vocational School, the neighbouring East Havana, Alamar and Villa Panamericana, the conglomerate formed by Lenin Park and the Zoological and Botanical Gardens. From 1 January 1959, Havana has provided an abundance of activities of all types that have constantly filled the streets, galleries, portals, avenues, sea walls, plazas, parks, even the *terrain vague* (Solà-Morales, 1995) such as the vacant and empty lots in neighbourhoods that have been used for impromptu political meetings or converted into outdoor television parlours. All this has constituted a symbolic appropriation of the city that was being reshaped by the elimination of private ownership of land, converting all the urban space into public space.

The Master Plan of 1970 – elaborated with the collaboration of the Cuban Mario González and Eusebio Azcue, the French Jean Pierre Garnier and the Italian Vittorio Garatti – proposed the elimination of walls, fences, and bars surrounding the extensive gardens of the mansions that belonged to the middle class in Vedado. This allowed pedestrians to reclaim these green spaces which until then had been off limits to the general public. This possession of the exclusive areas of the city by the people came about as a result of the mass exodus from the city of the wealthiest social strata.[8]

Thus, a spontaneous and festive social dynamic arose, albeit institutionally conditioned. Hence, if in the colonial and republican eras the iconic spaces of the city were isolated, over the last four decades they have been repossessed by the proletariat. A symbolic identification with places, whether close or removed, and whether architecturally lasting and monumental or passing and incidental, has also taken place. In light of a political climate moved by unpredictable and unexpected situations, the popular reactions in the city were also unusual and surprising. For instance, the empty expanse of the Civic Square was converted into the Revolution Square in the span of forty years, and has become the place for mass meetings of many thousands of residents and the nexus between the population and the political leaders. Unimportant buildings or spaces were also elevated to the category of urban symbols. Among these are the Embassy of Peru

Figure 9.4. Mass concentration in Revolution Plaza, *c.* 1960s. (*Photo*: Alberto Korda; Roberto Segre's archive)

Figure 9.5. Mass concentration in Revolution Plaza, *c.* 1970s. (*Photo*: Roberto Segre)

on Quinta Avenida (Fifth Avenue) of Miramar neighbourhood, where in 1980 10,000 people gathered and requested departure from Cuba. The building was converted into a museum and then demolished. In 2002 a statue of John Lennon was erected in an anonymous park in Vedado and is now a place of pilgrimage for foreign visitors and local youth.

Cuba's political system supposes participation in all officially programmed activities and, in some cases, activities that arise from popular and local initiatives. Thus, at the same time as the events that have become the staple of the main symbolic spaces of the city, the relative importance of the neighbourhood has taken on particular significance due to the increased political activism of the community. This was non-existent in the 1960s. It would have been impossible to think in the 1950s about collective street activities in upper- or lower-class neighbourhoods such as Víbora, Vedado or Miramar. Because of the scarcity of urban plazas, the street took the spotlight in public activities, integrated in the so-called 'Plan de la Calle'. Its importance was reinforced in a period of frequent electric blackouts, as neighbours gathered under the streetlights. With the organization of the Committees for the Defence of the Revolution (CDR), neighbourhood streets became the stage for political activity, anniversary festivals of the CDR, New Year's celebrations, activities for children organized during the Christian Holy Week – an effort to counterbalance the influence of the church over adolescents – and carnivals, dances and parades. These were initiatives that engaged the residents in activities in their neighbourhood spaces and offered entertainment animated by local cultural themes. As the streets and the façades of the houses were adorned with flowers, stars made of brass, paper flags and improvised posters of revolutionary, religious and syncretistic images, each neighbourhood assumed a different and renewed festive image that fleetingly transformed the perception of the city.

Some key sites established the iconic system of the Revolution in the urban context. These included the Presidential Palace, main headquarters of the government, whose ample plaza, Missions Avenue, set the stage for the first political gatherings; Revolution Square, whose space is dominated by the monument to José Martí and is equipped with a marble podium which constitutes a platform for politicians, foreign guests, and particularly for the speeches of Fidel Castro; the front steps of the university that are linked to students' struggles against the dictatorship, and are used as a stage for important youth activities; the Malecón, a privileged site for government organized protest marches due to its extension and spaciousness and the presence of the US Interests Section (the old Embassy). Some streets and avenues have also developed particular symbolic meanings. For example, San Lázaro Street, which extends from the university front steps, was the stage for student marches in honour of José Martí and of the medical students who were gunned down at the end of the nineteenth century; San Rafael Street, a classic thoroughfare of the business district that was redesigned and renamed El Bulevar and became an extension of the art gallery of Casa de las Américas (1971). Although it was lined with stores, at that time these lacked any products that could be sold. 23rd Street or La Rampa constituted the main artery of circulation alongside Vedado. The platform for the ceremony to honour the casualties in the explosion to sabotage the French ship La Coubre, which was transporting weapons to Cuba (1960), was built in front of Hotel Habana Libre (formerly the Havana Hilton). It was here that Alberto Korda took the famous photograph of Ché Guevara, which today serves as the iconic image of the guerrilla commander (Loviny, 2002). On that same road, in April 1961, on the corner of 23rd and 12th, not far from the cemetery, Fidel bid the victims of the mercenary airport bombings farewell, announced the attack on Playa Girón, and declared the socialist character of the revolution. In semantic terms, an area characterized by hotels and luxurious offices and nightclubs was transformed and

Figure 9.6. Pro-revolution march in the Havana Malecón, *c.* 1980s. (*Source*: Roberto Segre's archive)

assumed new uses and meanings. To this end, their introverted functions were replaced by the extroversion of public life.

When the Congress for the International Alliance of Architects (UIA), which drew thousands of professionals from all over the world, convened at La Rampa in 1963, it consolidated the place's recognition as a top recreational and cultural centre. In addition to the carnival dancers that travelled down 23rd Street for the festival's inauguration day, a covered 'salon' was built to house expositions – the Cuba Pavilion designed by Juan Campos – which from then on housed important national and international expositions (Segre, 1987).[9] In 1967, the park, which extended a whole block in front of Hotel Habana Libre, was covered with the large dome of the popular ice-cream parlour, Coppelia – designed by Mario Girona – intended to house hundreds of consumers at the same time (Sorkin, 2004). Another brief transformation of the street took place during the celebration of the centennial of Lenin's birthday in 1970. At that time graphic artists covered windows and façades with images of the October Revolution, figuratively transforming La Rampa into a street of Majkovski's Moscu.

Over the span of four decades, a system was born in which social and individual life was totally conditioned by politics, economy, and ideology. At the same time, a great part of these transcendental events were associated with the persistent conflict with the United States. Therefore, marches, mobilizations, parades, military exercises, and acts of defiance, defined the rhythm of the urban social life. Once Batista's dictatorship ended, the masses flooded the streets and seized the public spaces. The first violent reaction, which was short lived, was the destruction of the parking meters and slot machines downtown. With the arrival of Fidel Castro and the guerrilla commandos in Havana on 8 January 1959, the popular reception along the Malecón and the access roads to Columbia Barracks was tremendous. Alejo Carpentier describes this moment in the following:

I have never seen happiness, the happiness of the masses, the collective joy, that filled it today... The street, the public plaza, the main square, offer such a spectacle of enthusiasm, faith, participation in a great event, that I have never seen in Havana... And soon, 50,000 peasants mounting horses, with their hats made of palm leaf, their guayaberas, leather shoes, knives and machetes, they will march – Oh, Manes del Cucalambe! – on the streets of this joyous Havana of 1959. (Carpentier, 1959, 1996)

On 26 July 1959, it was the farmers, headed by Camilo Cienfuegos, that came from all the distant regions of the island and paraded in Revolution Square celebrating the anniversary of the assault on the Moncada barracks in Santiago, Cuba (1953). This image was recorded by Alberto Korda in the retrospective picture, 'The Quijote of Havana'.

It would be impossible to recount all the political events that marked the dynamics of urban life during this period. Between 1959 and 1961, some of the extraordinary events that took place in front of the Presidential Palace include Fidel Castro's first speech in front of a gathered crowd (that later supported him in the political crisis that motivated his momentary resignation as Prime Minister); the creation of the National Revolutionary Militias and the Committees for the Defence of the Revolution and Camilo Cienfuegos's last speech on 26 October 1959 prior to his disappearance in the ocean. However, the great empty space of the Civic Square was the ideal site for the great mass events. Renamed Revolution Square and framed by José Martí's monument and the inaccessible Palace of Justice, which was later adapted to house the headquarters of the government and Cuba's Communist Party, the square was the site of the main political, military, commemorative, social and cultural activities celebrated from 1 May 1959, when the Popular Militias marched for the first time. It was also the site of the tumultuous Catholic mass celebrated by the revolutionary priest Guillermo Sardinas in November of the same year. There were the set celebrations – the military march of 2 January, the popular march of 1 May (International Labour Day), the anniversary of the assault on the Moncada Barracks, the 26 July and the special celebrations, all of which usually gather a crowd of a million people. The most memorable ones were the popular approval of the Second Declaration of Havana against the sanctions imposed by the Organization of American States in 1960; the declaration of Cuba 'Territory Free of Illiteracy' in 1961; the solemn recognition of Ché Guevara's death in Bolivia in 1967, which was epitomized by the profound silence of the million people that stood at the plaza during the entire ceremony; the repeated silence to express the profound popular anger during the memorial for the athletes who were killed in the sabotage of the Cubana de Aviación aircraft off the coast of Barbados (1976) and for the Cubans who were killed during the United States' invasion of Granada (1983).

Not all the events celebrated in the plaza, however, had a serious or tragic character, and many were of a cultural nature. Take for example the chess games dedicated to adolescents and young people – the most recent one celebrated in 2002; the giant dinner in anticipation of the New Year in 1966 and 2000; the 1997 fair with multiple cultural activities or the symphonic concert with the participation of Silvio Rodríguez and Leo Brower in July 2004. Although in these events urban or architectural modifications of the plaza were not attempted, the buildings would be covered with graphic designs with a revolutionary context or slogans alluding to the ceremony. Hence with the celebration of the new anniversary of the birth of Ché Guevara in July 1988, graphic expressions were requested from artists and students from the national schools of art. At this time the word 'meditate' was unexpectedly written at the base of the monument, and quickly erased the day after the ceremony (Pedrosa Morgado, 2001). This was a critical gesture against the banal representations of the 'socialist realism' that dominated the graphic system at that time. Ten years later, in January 1998, the ideological changes in the government were evident. The state became more tolerant of religious expression, as when Pope Juan Pablo II visited Cuba and celebrated a mass at the plaza: although the image of Ché Guevara was still on the façade of one of the buildings, the façade of the National Library was covered with the figure of Jesus Christ and the words 'I believe in you' (Codrescu, 2001).

The festive character of the Malecón was never lost, and it remains a popular nightlife spot for lovers seeking a romantic rendezvous and a cool sea breeze. In 1978, in celebration of the Eleventh World Festival of Youth and Students, a plaza was inaugurated as a symbol of the event – albeit with limited urban and architectural visibility – in one of the existing spaces between the apartment

Figure 9.7. Revolution Plaza during the visit of John Paul II, January 1998. (*Photo*: Claudio Vekstein)

buildings of the 1950s. Later, during the month of July and throughout the 1980s, the Malecón became a stage for parading exotic and monumental carnival floats. This event would usually pull together all of Havana. The Malecón's first political use was in the annual visit of the primary school students who would toss flowers into the sea, on the anniversary of the disappearance of Camilo Cienfuegos. The frightening images of anti-aircraft machine guns located along the coast, as it was during the stressful days of the October Crisis (1962), are still remembered.

The headquarters of the US Interests Section became the site of massive protests every time a hostile act was exercised against Cuba. On several occasions since 1964, Cuban fishermen have been illegally detained at sea, giving rise to an immediate people's reaction at the Malecón. At the height of tensions in the 1990s, several mass marches took place along the coast. The climax, when Elián González – a child castaway and survivor of a raft that was fleeing the island who was retained in Miami in 1999 – generated daily mass demonstrations on the steps of the US Interests Section building demanding his return to the country. Such occurrences prompted the construction of a metal structure of poor design. Officially known as the Anti-Imperialist Tribune and commonly referred to as 'the protest-dome', it is adorned with a 'socialist realist' image of José Martí, and has disfigured the impeccable form of a building reflective of the fine architecture of the 1950s. Perhaps a more fortunate presence in the plaza would have been the monument designed by Oscar Niemeyer in the late 1980s, which is a metaphorical image of the break up of the US embargo over the island.

In the 1990s, with the crumbling of the socialist world and the disintegration of the USSR, the island was beset by serious and dramatic financial troubles. This was caused by the interruption of petroleum supply and the business nexus with eastern European countries. Urban life also suffered direct consequences. This was not only due to the insurmountable difficulties caused by the public transport crisis, the lack of electricity and the halt in construction, but also to the loss of the social cohesion characteristic of the initial stage of the Revolution. The individual and persistent daily struggle for material survival weighed heavily on any identification with the unattainable social and political utopia.

With the acceleration of tourism, the presence of foreign companies and the free circulation of the dollar prompted by the financial crisis, the urban area started to fragment and differentiate in its use as well as in its architectural quality. The buildings constructed in the residential neighbourhood of Miramar – hotels and apartments for foreign travellers – assumed a kitsch flavour similar to urban speculative initiatives of capitalist countries (Coyula, 2001; Scarpaci, Segre and Coyula, 2002). This curtailed the environmentally homogenous character of the area dating from the 1950s. At the same time, the old urban inequalities seemed to return: Old Havana was partially restored for the privilege of tourists, although it was furnished with social infrastructure for the local population; new hotels and shopping centres of dubious architectural quality in Vedado emerged; and the residential neighbourhood of Miramar, which was used for office development such as the Miramar Trade Centre, was dotted with stores and banal supermarkets patronized by those with access to US dollars.

With the decline in daily political activity, business trade became the city's main engine of social life, redefining the symbolic contents of public places. The rapid deterioration of popular neighbourhoods continued due to lack of maintenance of streets and buildings. The ruinous state of downtown Havana became a recurrent theme in the literary works of Pedro Juan Gutiérrez (1998) and Abilio Estévez (2004) and such films as Win Wenders's *Buena Vista Social Club*, Fernando Pérez's *Suite Havana* and the recent (2005) Spanish-Cuban film, *Havana Blues* – scrutinizing the dissimilarity of inherited 'noble' areas of the city. The local people concentrated themselves in spaces adapted for business activities, typically of the small private vendor type, in temporary and permanent kiosks in pursuit of the mighty dollar. Artisans have also set up improvised stands and made use of plazas, streets, portals and vacant lots. These are distributed all over the city but are mainly located along the Malecón and in the tourist areas surrounding the historic centre.

At the dawn of this century Havana will have existed for half a millennium, inheriting spaces of extraordinary urban and architectural value. The beautiful plazas of the colonial era, the tree shaded boulevards representative of the first stages of the republic, and the vitality of Vedado built in the 1950s, made up the habitual public spaces where ordinary and extraordinary historical events took place (Coyula, 1996). However, a new dimension of 'socialist' space in the urban context that could have expanded the whole city's aesthetic values and symbolic meaning did not materialize. It is not by chance that the only two buildings important enough to house public life – the Cuban Pavilion and Coppelia – were built in La Rampa. The rest of the great works of the Revolution were isolated in the urban periphery.

In recent decades, Havana has suffered from the lack of attention needed to sustain the daily life of its neighbourhoods and to regenerate the cultural and economic development that existed in some of them: Víbora, Lawton, Cotorro, and Marianao. Abandoned to the interests of private initiatives in the face of the Revolution, and now marginal to state policy, these neighbourhoods would have benefited from the creation of new centres of social life, in which a mixture of functions would have permitted a new metropolitan dynamism in an increasing new polycentric capital city. If the city is a material reflection of the dynamics established by social life, in the dialectic of the private and public initiative, Havana today reflects the existing tensions of a political system that boldly struggles for its survival in a hostile and aggressive world. It is from this point that few events attract a people whose daily struggle for financial reward leaves them with no other choice but to retreat from public life. This is a situation that is reflected in the deterioration of various urban public places. The fading of the ideological euphoria that filled the streets and plazas of the Revolution has not been substituted by democratic participation of residents and urban planners, and remains in the grip of the central government. However, a desire remains in many for a future without the banal, imperial images that rule in capitalist cities, already being imported to Havana by real estate, tourism and consumer shopping interests. The old and noble Havana, a venerable place almost 500 years old, anxiously awaits the promise of better times.

Notes

1.   The term *criollo* or *criollos* refers to Latin Americans of European parentage, typically Spanish parentage.
2.   It has been widely documented that the first cities were not related to the Renaissance urban tradition but with the rules of the Laws of the Indies (Chueca Goitía and Torres Balbás, 1951; Martínez Lemoine, 1977).
3.   Alejo Carpentier was able to describe the excitement of the harbour, the coming and going of travellers, its colours, smells, and music, framed by the subtle baroque style of its architecture – attributes that gained the city the name 'The Pearl of the Antilles' (Carpentier, 1962).
4.   In 1895, for instance it became a rallying symbol for the efforts of Spanish volunteers who defended Spain against the Cuban patriots; in 1898 few attended the departure of the defeated Spanish volunteers who returned to Spain.
5.   'Concentration' was a measure taken by Weyler, who expelled the farmers from their lands relocating them in the suburb adjacent to the city with the intention of weakening the Ejército Libertador, which was in the Havana region. As consequence of poor hygienic conditions they were decimated by epidemics, especially by yellow fever.
6.   In 1958 the Malecón achieved world-wide fame when the Argentine pilot, Juan Manuel Fangio, was kidnapped by Fidel Castro's urban guerrillas.
7.   Undoubtedly this was more favourable to the celebration of massive gatherings in later years.
8.   In 1960, for instance, the illiterate children of peasants from western provinces were taken to Havana to study. They were housed in the empty residences of Miramar. Phenomena such as this did not take place to the same degree in Santiago, Chile during the government of Salvador Allende, and is not currently happening in Caracas with Hugo Chávez. These are cities in which the entrenchment of the bourgeoisie in their neighbourhoods has generated a political duel in the urban space, among antagonistic social groups (García-Guadilla, 2003; Irazábal and Foley in this book).
9.   The exhibition of the Paris May Saloon works of art at the Cuba Pavilion in 1967 provided the Cuban population with an unusual experience (Rigol and Rojas, 2004).

References

Álvarez-Tabío Albo, Emma (2000) *Invención de La Habana*. Barcelona: Editorial Casiopea.
Baroni Bassoni, Sergio (2003) *Hacia una cultura del territorio*. Havana: Grupo para el Desarrollo Integral de la Capital (GDIC).
Bretos, Miguel A. (1996) Imaging Cuba under the American Flag: Charles Eduard Doty

in Havana, 1899–1902. *Journal of Decorative and Propaganda Arts*, No. 22 (Cuba Theme Issue), pp. 83–103.

Cabrera Infante, Guillermo (1987) *Tres tristes tigres*. Barcelona: Seix Barral.

Carpentier, Alejo (1962) *El Siglo de las Luces*. México: Compañía General de Ediciones.

Carpentier, Alejo (1959, 1996) Una jubilosa Habana (cited in *El Nacional*, Caracas, 17 June 1959), in Alejo Carpentier (1966) *El amor a la ciudad*. Madrid: Alfaguara, pp. 93–96.

Chacón Nardi, Rafaela (2002) Adoro mi ciudad, in Coyula, Mario (ed.) *La Habana que va conmigo*. Havana: Editorial Letras Cubanas, pp. 36–47.

Chateloin, Felicia (1989) *La Habana de Tacón*. Havana: Editorial Letras Cubanas.

Chueca Goitía, Fernando, Torres Balbás, Leopoldo (1951) *Planos de ciudades iberoamericanas y filipinas existentes en el Archivo de Indias*. Madrid: Instituto de Estudios de Administración Local.

Codrescu, Andrei (2001) *Ay Cuba! A Socio-Erotic Journey*. New York, NY: Picador.

Coyula, Mario (1996) La Habana siempre; siempre mi Habana. *Archivos de Arquitectura Antillana*, No. 2, pp. 77–83.

Coyula, Mario (2001) Playa frente al espejo: desafíos del XXI. *Revolución y Cultura*, No 2, pp. 4–18.

Estévez, Abilio (2004) *Os Palácios Distantes*. São Paulo: Editora Globo.

Evans, Walker (2001) *Cuba*. Los Angeles, CA: The J. Paul Getty Museum.

García-Guadilla, María Pilar (2003) Territorialización de los conflictos-sociopolíticos en una ciudad sitiada: *guetos* y feudos en Caracas. *Ciudad y Territorio. Estudios Territoriales*, **35**(136–137), pp. 421–440.

Guimerá, Agustín and Monge, Fernando (eds.) (2000) *La Habana puerto colonial (siglos XVIII–XIX)*. Madrid: Fundación Portuaria.

Gutiérrez Viñuales, Rodrigo (2004), *Monumento conmemorativo y espacio público en Iberoamérica*. Madrid: Ediciones Cátedra.

Gutiérrez, Pedro Juan (1998) *Trilogía sucia de La Habana*. Barcelona: Anagrama.

Hardoy, Jorge E. and Aranovich, Carmen (1969) Urbanización en América Hispánica entre 1580 y 1630. *Boletín del Centro de Investigaciones Históricas y Estéticas*, No.11. Caracas: Universidad Central de Venezuela, Facultad de Arquitectura y Urbanismo, pp. 9–89.

Izquierdo, Yolanda (2002) *Acoso y ocaso de una ciudad. La Habana de Alejo Carpentier y Guillermo Cabrera Infante*. San Juan: Isla Negra Editores, Universidad de Puerto Rico.

Lejeune, Jean François (1996) The city as landscape: Jean Claude Nicolas Forestier and the great urban works of Havana, 1925–1930. *Journal of Decorative and Propaganda Arts*, No. 22 (Cuba Theme Issue), pp. 151–185.

Lejeune, Jean François (ed.) (2003) *Cruauté & utopie. Villes et paysages d'Amerique Latine*. Bruxelles: Centre International pour la Ville, l'Architecture et le Paysage, CIVA.

Le Riverend Brusone, Julio (1992) *La Habana. Espacio y vida*. Madrid: Editorial Mapfre.

López Segrera, Francisco (1989) *Cuba: cultura y sociedad (1510–1985)*. Havana: Editorial Letras Cubanas.

Loviny, Christophe (ed.) (2002) *Cuba par Korda*. Paris: Calmann-Lévy, Jazz Éditions.

Martínez Lemoine, René (1977) *El modelo clásico de ciudad colonial hispanoamericana*. Santiago de Chile: Facultad de Arquitectura y Urbanismo, Universidad de Chile.

Ortiz, Fernando (1943) *La hija cubana del Iluminismo*. Havana: Molina & Cía.

Ortiz, Fernando (1963) *Contrapunteo cubano del tabaco y el azúcar*. Havana: Consejo Nacional de Cultura.

Pedrosa Morgado, Concepción (2001) *Imago Ilha: Epifanía da 'Generación de los Ochenta' cubana*. Niterói: Universidade Federal Fluminense, Instituto de Arte e Comunicação Social.

Pogolotti, Marcelo (2002) *La República de Cuba a través de sus escritores*. Havana, Editorial de Letras Cubanas.

Portuondo, José Antonio (1962) *Bosquejo histórico de las letras cubanas*. Havana: Editora del Ministerio de Educación.

Raffy, Serge (2003) *Castro l'infidèle*. Paris: Librairie Arthème Fayard.

Rigol, Isabel and Rojas, Ângela (2004) Entre nostalgie et sauvegarde: les hauts et les bas de La Rampa. *L'Architecture d'Aujourd'hui*, No. 350, pp. 54–61.

Rodríguez, Eduardo Luis (1998) *La Habana. Arquitectura del siglo XX*. Barcelona: Blume.
Rodríguez, Eduardo Luis (1999) *The Havana Guide. Modern Architecture 1925–1965*. New York, NY: Princeton Architectural Press.
Roig De Leuchsenring, Emilio (1964) *La Habana. Apuntes Históricos*. Tomos I-II-III. Havana: Oficina del Historiador de la Ciudad de La Habana, Editora del Consejo Nacional de Cultura.
Rojas Mix, Miguel (1978) *La Plaza Mayor*. Barcelona: Muchnik.
Rovira, Jose M. Rovira (2000) *José Luis Sert 1901–1983*. Milan: Electa Editrice.
Scarpaci, Joseph; Segre, Roberto; Coyula, Mario (2002) *Habana. Two Faces of the Antillean Metropolis*. Chapel Hill, NC: The University of North Carolina Press.
Schwartz, Rosalie (1997) *Pleasure Islands. Tourism and Temptation in Cuba*. Lincoln, Nebraska: University of Nebraska Press.
Segre, Roberto (1995) La Habana de Sert: CIAM, Ron y Cha-Cha-Cha. *Documentos de Arquitectura Nacional y Americana*, Nos. 37/38, pp. 120–124.
Segre, Roberto (1996) *La Plaza de Armas de La Habana. Sinfonía urbana inconclusa*. Havana: Editorial de Arte y Literatura.
Segre, Roberto (1987) *Arquitetura e Urbanismo da Revolução Cubana*. São Paulo: Editora Nobel.
Segre, Roberto (2002) Havana, from Tacón to Forestier, in Arturo Almandoz (ed.) *Planning Latin America's Capital Cities 1850–1950*. London: Routledge, pp. 193–213.
Solà-Morales Rubió, Ignasi de (1995) *Terrain Vague*, in Davidson, Cynthia (ed.) *Anytime*. Cambridge, MA: MIT Press.
Sorkin, Michael (2004) Let a thousand flowers bloom: Cuban Modernism's short moment in the revolution's sun. *Architectural Record*, No. 12, pp. 67–69.
Terán, Fernando (1989) *La ciudad hispanoamericana. El sueño de un orden*. Madrid: Centro de Estudios Históricos de Obras Públicas y Urbanismo, Ministerio de Obras Públicas y Urbanismo.
Thomas, Hugh (1998) *Cuba or The Pursuit of Freedom*. New York: Da Capo Press.
Venegas Fornias, Carlos and Núñez Jiménez, Antonio (1986) *La Habana*. Madrid: Instituto de Cooperación Iberoamericana.
Venegas Fornias, Carlos (1990) *La Urbanización de Las Murallas. Dependencia y Modernidad*. Havana: Editorial Letras Cubanas.
Venegas Fornias, Carlos (2003) *Plazas de Intramuros*. Havana: Consejo Nacional de Patrimonio Cultural.
Vidler, Anthony (2004) Redefinido a esfera pública. *Jornal Arquitetos*, No. 215, pp. 53–63.

Acknowledgements

I would like to thank Mario Coyula and Eliana Cárdenas, Professors at the School of Architecture at the Polytechnic Institute 'José Antonio Echeverría', Havana, for their contribution to facts used in this chapter; Concepción Pedrosa, Emma Álvarez Tabío, José Prado, Ramzi Farhat, and Clara Irazábal for their assistance and comments throughout the process of editing this text.

Chapter 10

Unresolved Public Expressions of Anti-Trujilloism in Santo Domingo

Robert Alexander González

In 1961, the reign of Rafael Leonidas Trujillo y Molina (1891–1961) in the Caribbean nation of the Dominican Republic, one of the most enduring and bloodiest dictatorships of the twentieth century, finally came to an end.[1] In the aftermath of his demise, traces of Trujillo y Molina's totalitarian regime, in particular monuments glorifying his life and works, have been systematically removed. In the subsequent restructuring of the urban landscape, an interesting question is whether and how Dominicans have, in a new era of democracy, attempted to represent their new political order in the public realm, compared with how this representation was managed by the previous regime. As a corollary question, how were the extraordinary events of the Trujillo years subsequently processed through the few sites of memory that have been built since throughout the city?

Santo Domingo's linear urban pattern, decentralized system of public spaces, and monument building practices contributed to the formation of democracy and citizenship in ways that are unique among Latin American cities. The evidence can be observed in the study of key memorials throughout Santo Domingo and in press accounts published during the city's renaissance and reinvention after Trujillo's demise. Admittedly, Dominicans have made use of public spaces in Santo Domingo's colonial centre throughout its long history in a fashion that is typical of Latin American cities. Central colonial spaces, representing both a colonial heritage and often, local or national administrations, serve as the appropriate background for most public events and celebrations. During Trujillo's reign, for example, the Puerta de San Diego (or del Conde), also called the *'altar de la Patria'* (altar of the homeland), was a focus in the centenary celebration of

Figure 10.1. The military procession taking the remains of the founding fathers to the Conde de San Diego. (*Source*: *Album del Centenario*, 1944)

1944. The country's founding fathers, after their exhumation, were re-interred here during the celebration (figure 10.1). For Trujillo, this event served to further elevate his stature by allowing him to create a symbolic link between himself and the country's glorious past. This was also the place where the masses congregated on 2 September 1961, after Trujillo's fall, an event that celebrated the revocation of the Trujillo regime and the subsequent re-establishment of democracy. Thus the Puerta de San Diego served as a space for both the centenary event and the demonstrations celebrating the fall of the Trujillo regime, but photographic records of the two events reveal the difference between the enforced public order in the former, and public freedom of expression in the latter. Trujillo, during his reign, had also used newly created public spaces to deploy and reflect the totalitarian regime. He memorialized his own regime with the design of a set of plazas in the western part of the city, dedicated at the *Feria de la Paz y Confraternidad del Mundo Libre* (The Peace and Brotherhood in the Free World Fair) in 1955–1956 (figure 10.2). In addition to these public spaces, the city has others also suitable for mass gatherings. In fact, the absence of a central space that singularly represents the Dominican Republic's body politic, like the Zócalo in Mexico City, dissipates public activity in Santo Domingo over numerous plazas. These public spaces were created and appropriated by those whose intentions were to politicise urban space as well as merely improving traffic flow or developing neighbourhoods.

As the totalitarian regime unravelled in 1961, it was finally possible for Dominicans to imbue their urban spaces with new meaning, heeding the calls of the partisan newspaper *Unión Cívica*, which actively encouraged the assertion of democracy and citizenship in its weekly columns.[2] While these newspaper

Figure 10.2. Aerial view of the *Feria de la Paz*. (*Source*: Tourist brochure)

columns served as bases for the exploration of democracy and citizenship by the local population, with their lively debates in the form of essays and public lectures, democracy and citizenship were enacted in physical space mainly through the act of erasure. Immediately after Trujillo's demise, the Dominicans responded with inaction when it came to the physical environment. The project of monument building that reflected critically upon the Trujillo era would not occur until much later, and questions still remain as to the extent to which Dominicans drew from their collective memory and how public participation played a part in this process. To attract visitors and provide for the expectations of tourists, the choice was also made to de-emphasize evidence of the Trujillo era in the public realm. This can be seen from a review of various guidebooks, tourist maps, and city guides from different periods. Yet, because Trujillo's imprint on the urban environment was so extensive, it is inevitable that his mark, evident in the large-scale monuments and urban renewal projects that he directed, was substantial. Consequently, his mark remained at the centre of many urban experiences, even when tourists were not aware that they were amidst interventions of his legacy. The resulting dual urban imaginary pits an oppressed set of representations, easily identified by the local community and supported by oral histories (i.e. sites of atrocities, of past memorials and of acts of homage to Trujillo), against a set of public representations, aiming to attract tourists and supported by city generated propaganda (often erasing Trujillo from the picture).

Santo Domingo's urban fabric consists of interconnected nodes of decentralized public spaces that stretch westward along the *malecón*, the coastal road named Avenida George Washington, and extends across the Ozama River to the eastern sector where the famous Columbus Memorial Lighthouse stands. The linear nature of the city means that there is no single point of convergence. This may have hindered the efficacy of public protests throughout Santo Domingo's long history, and may have also diminished the symbolic importance of any one monument in the post-Trujillo years. The linear nature of the city is the result of largely westward expansion from the colonial centre. Unlike most other colonial towns, Santo Domingo did not grow outwards from one single public space. Historian Erwin Walter Palm observed that, since the turn of the sixteenth century, the Parque de Colón, with its *cabildo* and prison, had coexisted with a second plaza, where the Alcazar de Colón (Plaza de Armas and the Palacio de Capitanes Generales) was located (Palm, 1974). This urban bipolarity was abolished with the creation of a third major public space, the Plaza de Independencia, in the late nineteenth century, featuring statues of three Dominican heroes who fought for independence from Haiti in 1844. In the De Moya map, dated 1900, we see a record of the westward and northward expansion of the city. The new development west of the historic sector, identified as *Ciudad Nueva* in this map, was the work of the engineer J.M. Castillo, and it corresponded with the end of the dictatorship of Ulises Heureaux (1882–1899), who had set these plans for city expansion in motion. By this time the major streets had been extended beyond the colonial centre, and in doing so the city's historic walls were destroyed (Pérez Montás, 1998). Development under Trujillo continued along this trajectory by extending the *malecón* towards Güibia beach, then in 1955 to the site of the *Feria de la Paz*, and eventually all the way to Bajos de Haina. Armed with this urban history, we can appreciate the importance of the city's many public spaces and urban icons, particularly as they provide an episodic narrative of the city's development.

The theme of extraordinary events in ordinary spaces is of particular importance in the case of Santo Domingo, in light of its legacy of totalitarian rule and its particular pattern of city growth. If the other case studies presented in this volume illustrate the extent to which extraordinary events are manifest in public spaces in any specific historic episode, Santo Domingo illustrates, on the contrary, the consummate form of absence. When thinking of the most memorable events of public expression in ordinary places, visions of masses of people claiming highways, thoroughfares, and the most public of civic spaces come to mind. In Latin America, we have seen how, time and time again, this type of *ad hoc* disorderliness has aided the masses in reclaiming powers from the state they thought it had illegally confiscated. However, in the public spaces associated with the Feria de la Paz in Santo Domingo, which are remnants of the extraordinary events imposed by the totalitarian regime, we see that a civic icon of the past can easily exist as a denigrated urban void. The abandonment was such that recently an urban design competition was launched to address its ineffectiveness as a public space. While the sort of events explored in this volume did occur briefly in

1961 with Trujillo's fall, marking the act of erasure as one of the most interesting extraordinary events in the city, the weight of the nation's totalitarian legacy has seemingly chipped away at the political agency of the public, paralysing it and preventing it from politicising the public realm. The extraordinary events of the past have resulted in a cavalier attitude towards the urban landscapes of the dictatorship, and the response has been to downplay the importance of these types of events altogether.

This chapter begins with the methods used by Trujillo to transform his country into a self-referential shrine, beginning with his attempts to build the Columbus Memorial Lighthouse early in his dictatorship and ending with the Feria de la Paz and his eventual fall. Trujillo's self-idolization through restructuring the urban imaginary, which occurred between 1930 and 1961, is presented here to provide the context for understanding how the eventual erasure of signs of authoritarianism took place in physical space. A majority of this activity was captured in local Dominican newspapers, and *Unión Cívica* serves as an excellent window onto this extraordinary period of erasure. A review of the subsequent period of monument building and rededication of major spaces in Santo Domingo, which takes us to the present, reveals the power of the transformation of space in the activation of political thought, but it also raises questions about the vulnerability of these spaces. The contrast between a more recent, subdued, and slow effort to memorialize the past and the active period of erasure captured in *Unión Cívica* is telling. The chapter ends with a brief consideration of an urban design competition, entitled 'Revitalization Plan for the Centre of the Heroes of Santo Domingo', held in 2005, which aimed at redefining citizenship and nationalism for the Dominicans, specifically by returning to the fairgrounds and with the aim of transforming these same spaces once again.

Memory Crisis and Official Histories

Memory plays an important role in the study of how Santo Domingo's public space has undergone constant transformation from the Trujillo era through the years following his fall from power up to the present day. Memory is understood here both as part of an 'official history', carefully crafted by the ruling party, and as part of the personal narratives which can challenge 'official histories'. An image which appeared in *Unión Cívica* in 1961, entitled '¡*Fuera Las 'T'*!' ('All 'T's' be gone!' referring to Trujillo), expressed the desperate need Dominicans felt to regain control of official memory, and the power of the exhortation to remove immediately all evidence of Trujillo from the public realm.[3]After this period, there was a time when some Dominicans were willing to entertain competing accounts of the Trujillo era. Only very recently, however, are carefully researched narratives of past extraordinary events beginning to emerge as part of the city's 'official' history. In this day and age, visitors and tourists expect to learn about a city's entire history, both the laudable and regrettable. Many cities present memorials to historic struggles prominently and proudly for visitors to see, for example the Plaza de Ttatelolco in Mexico City, Statue Park in Budapest, or the Civil Rights Museum in Memphis. However, the 'unresolved public expressions' of this chapter's title points to the fact that Santo Domingo is not yet willing to reveal to its visitors its full history as its 'official history'.

The interest in considering what a city looks like through the mediated lens constructed for the tourist stems from the desire to examine not just memory, but a 'memory crisis' that has influenced the design of our cities, and all too often narrows our understanding of them. Critic and philosopher Walter Benjamin considered a memory crisis the haunting preoccupation of nineteenth-century city development, noting that particularly in Europe, major and minor cities alike looked for any opportunity to capitalize on a good story framing the city history in a way that reflects well on the city. Benjamin believed that in cities in general 'the traditional sense of memory was forgotten and replaced by "official" history, which is connected to the pervasive existence of historic districts in our cities' (Boyer, 1998). How do we measure the influence this memory crisis has had on preservation movements, the writing of 'official' histories, or the multi-million dollar tourist industry? Boyer (1998) expands on the notion of memory crisis and terms the resultant dual mapping of cities 'the descriptive map' and 'the spectacular map'. The descriptive map is the systematic map that outlines the many sites a city offers a tourist for consumption. Accordingly, 'a universal methodology of guidebooks eventually is the result'. I would argue that in Santo Domingo, the descriptive map is still incomplete. The spectacular map is less objective and emphasizes the grand and visionary aspects of a city; it emphasizes 'wondrous landscapes, the grandeur of architectural marvels, the curious customs of peoples and habits of place'. These maps could provide a complete picture of the city imaginary, but, obviously, only when they are considered together. In Santo Domingo, I argue that the spectacular map is the distraction away from the horrific legacy of the past.

Santo Domingo is 'spectacularly' presented to the tourist using a rhetoric which prominently features its colonial sector, and extols its being the first settlement in the Americas. This positions the city of Santo Domingo as the 'first' city – the site of the first Columbus landing. The local and national aim of this rhetoric is to establish the Dominican Republic as the birthplace of the New World, and Santo Domingo as the repository of many institutional 'firsts' in the Americas (i.e. the first cathedral of America, the first College Gorjón, the first hospital, etc.). Because Trujillo's most prominent interventions are located in the western part of the city, it should be noted that they are missed as many tourist guides only present zoom-in maps of the colonial sector. The most emphatic expression of this rhetoric of 'firstness' is not found in the colonial sector of the city, however, but in the Columbus Memorial Lighthouse, the peculiar structure that sits across the Ozama River. Having been in the planning stage for over 60 years and finally completed in 1992, this monument was meant to celebrate the Western Hemisphere's 'discovery' by Columbus, but many locals would argue that it only represents the most wasteful effort to assert a position of primacy for the Dominican Republic.

Just as many Dominicans would like to move beyond the focus on Columbus's 'discovery' as the defining point of their history, many would also like to move beyond the legacy of Trujillo. Dominican politicians, historians and civil boosters are still actively redefining their cultural narratives, and collective memory is carefully reconsidered as contemporary journalists, historians, writers, and artists negotiate sensitive issues of the past. The recent urban design competition, aimed at refashioning the fairgrounds of the Feria de la Paz was a key opportunity to revisit this desire to move on. Issues of citizenship, specifically how 'Dominicaness' should be represented in the public realm, did play a central part in this competition, evident in the competition requirements and also reflected in some of the entries. Yet as long as the public realm bears any trace of the Trujillo era, whether as old sites transformed for new purposes or as new sites referencing unfortunate bygone eras, they remain contentious elements of the urban fabric. The deep emotional wounds of the Trujillo era understandably have left visible, physical scars. In coming to terms with their history, Dominicans struggle with whether to reveal the scars and suffer the effects of remembering or to erase the scars and subvert the memory.

Transforming the Nation into an Auto-Shrine

Trujillo's ascendancy to power was made possible by his domination of military power and the continuous intimidation of the country's citizens. In the public realm, Trujillo's rise to power was propelled by staged political rallies that increased in number with the emergence of the Dominican Party in 1931, which was superseded a decade later by the Trujillista Party (Atkins and Wilson, 1998, p. 65; Roorda, 1998, pp. 91–94). As historian Paul Roorda (1998, p. 95) has noted, 'The Trujillista ideology dictated that one of the main functions of public life was the perpetual deification of *El Jefe* ("The Chief", one of Trujillo's many monikers) through rhetorical acts of homage, nationalistic rituals, and the proliferation of images and place-names associated with him. The process by which Trujillo's name and image were conflated with the Dominican nation itself began in earnest on Dominican Independence Day, 27 February 1933, when the title "Benefactor of the Republic" was bestowed upon him'. Roorda notes how Trujillo was even publicly compared to Buddha, Confucius, Christ, and Mohammed. The large 'Civic Reviews' that his administration held drew tens of thousands of people who joined in the public ceremonies, many doing so out of fear, but also encouraged by the promise of free meals. For generations historians and tourist boards had glorified the role of Columbus and the founding fathers in making Santo Domingo and the Dominican Republic what it was. Trujillo must have had the ambition to join the ranks of the likes of Columbus in his work on the capital city and the nation. The so-claimed remains of Columbus and the founding fathers were ceremoniously interred in the public realm, and the question of where Trujillo would be laid to rest one day must have motivated his course of action.

The story of Trujillo's self-idolization begins in 1930, a year which could be seen as a turning point for the city of Santo Domingo. In May that year, Trujillo took office, and on 3 September, only days after his inaugural 'coronation' (Roorda, pp. 54–55), Hurricane San Zenón, like a premonition of future horrors, virtually levelled Santo Domingo. The extent of the destruction throughout the city conveniently provided Trujillo with an urban *tabula rasa*, allowing him to plan and rebuild in a way that reflected his taste and expressed his power. A significant opportunity for both monument and legacy building came his way only fourteen months after taking power with the Columbus Memorial Lighthouse. An international assembly of jurors, consisting of architects Eliel Saarinen (Europe), Horacio Acosta y Lara (Latin America), and Frank Lloyd Wright (North America), awarded first prize to a British architecture student, Joseph Lea Gleave, in Rio de Janeiro.[4] Thus, the history of civic improvements under Trujillo began with the Lighthouse as his first civic project, but he was unable to carry this to completion due to inadequate support from donor nations.[5] For many years, Trujillo earnestly engaged with the Washington DC-based Pan-American Union (today the Organization of American States), the sponsors and organizers of the project, in an unsuccessful effort to raise funds. The Pan-American Union terminated its activities in connection with the lighthouse in 1949, sending a check for $26,122.56 to the Dominican Republic representative in Washington.[6] Unsuccessful in his bid to complete the Lighthouse, Trujillo looked to other grand projects, as well as to the complete transformation of the capital city itself, to make his mark.

Every Dominican city experienced some degree of change under Trujillo, but none so much as Santo Domingo, which was officially renamed Ciudad Trujillo in 1936. The physical alterations to the city fabric were the most significant that the city had ever seen. As historian Frank Moya Pons points out:

... as soon as the economy began to recover around 1938, the government resumed the previous projects of highways, bridges, irrigation canals, and agricultural settlement with unprecedented energy ... the Dictator worked to transform the old city of Santo

Domingo into the principal industrial center of the Dominican Republic... (Moya Pons, 1995, p. 361)

Generalísimo Trujillo, as he was also often referred to, was called 'the patron saint of architecture'. Numerous publications of that era posited that his great infrastructure projects could be directly linked to the increasing national production of manufactured goods. One publication depicted the dictator wearing a suit, patterned with bricks, with his heliographic image surrounded by the many buildings and public projects that he realized (Pieiter, 1958; Walker, 1956).

The book, *La Arquitectura Dominicana en la Era de Trujillo* (1949) showcased the President's many accomplishments.[7] His interventions resulted in an eclectic collection of buildings that ranged in style from neo-classical to modern. Called *'El Primer Arquitecto de la Nación'* in the book's introduction, Trujillo's alliance with the architectural profession is made clear, especially his collaboration with the engineer-architect Henry Gazón Bona, who was also head of the national army. These visionary projects reflect an administration with stylistically diverse taste, supported by an army of engineer-architects, students, and city officials, all at Trujillo's beck and call.[8] The projects presented in the book include monuments, hotels, bungalows, and market places. In addition to these interventions, Trujillo has also been noted for initiating the preservation movement in Santo Domingo, which led to the rebuilding of the entire colonial quarter after Hurricane San Zenón.

Every new project built during this dictatorship literally bore the stamp of Trujillo. Historian Howard J. Wiarda (1970, p. 134) writes: 'All buildings constructed during the Trujillo era contained a cornerstone with his name engraved, and excerpts from his speeches were carved into the walls'. This form of deification was not limited to the urban areas:

Signs were placed throughout the Dominican countryside so that people could not forget the author of their benefits and good fortune. Signs in the hospitals read 'Only Trujillo Cures Us' and those at the village pumps 'Only Trujillo Gives Us Drink'. Other slogans such as 'Trujillo Is My Protector' and 'Trujillo Always' appeared in academic theses and on plaques hung on public buildings. (*Ibid.*, p. 135)

The national celebration held in 1944, the Centenary of the Dominican Republic, served as a template for the ever-grander events that followed, especially in the way Trujillo showcased himself and his commitment to public works projects. The *Album del Centenario de la República Dominicana* was produced to document this event, which took place from 22 February to 4 March. Thirty countries participated and their representatives witnessed the transformation of the capital city. Conferences, entertainment, and the dedication of numerous memorials were some of the activities that packed this 10-day event. A new airport and a racecourse were also inaugurated and, as the pinnacle event of a massive gathering, the founding fathers – Duarte, Sánchez and Mella – were re-interred together in the city's new public 'altar', the Puerta de San Diego. A foundation stone was set at the Columbus Memorial Lighthouse site, and the Director of the Pan-American Union, L.S. Rowe, delivered a public lecture. Celebrants danced late into the night at the new, modern hotels – the hotels Jaragua and Embajador.

The Feria de la Paz y Confraternidad del Mundo Libre, 1955–1956

Trujillo's grandest attempt to transform the urban fabric of Santo Domingo was realized in the Feria de la Paz y Confraternidad del Mundo Libre, which drew its inspiration from the United States, the country that he admired and that had shaped his military career – he was trained by the US Marine Corps. He borrowed easily recognizable iconography – the Washington Memorial, for example – and renamed the prominent *malécon*, Avenida George Washington, and another major thoroughfare, Avenida Abraham Lincoln. His grandiose fair, which was fully documented in two volumes of the *Albums de Oro*, however, was the pinnacle of this pervasive mimicry, making unmistakable references to New York's World's Fair of 1939. The *Albums* illustrate the fair's offerings, depicting the Trujillo family amidst the lavish fairgrounds. Never sanctioned by the Bureau of International Expositions in Paris, this was more of an industrial fair. Cucho Alvarez Pina, President of the Administrative Council of Ciudad Trujillo, had organized a competition for the design of the fairgrounds. The Civic Centre facilities, a group of institutional buildings, were to serve as a stage for exposition activities; these facilities housed state and national offices after the fair ended (Moré and Matinez, 2005, p. 97). The evolution of the fair's main plaza is an example of the transformation of civic space. It was originally designed to celebrate the 'great works' of Trujillo's regime, but was later used as a stage for protesting the human rights abuses of that dictatorship. Most recently, the 2005 competition proposed using the plaza to rebrand contemporary Dominican citizenship and nationalism.

The commission for the fair's final design was given to the Yale-trained architect Guillermo González Sánchez (1900–1970).[9] His design was in line with earlier interests in creating an entirely new civic centre, shifting the city's centre of gravity further west. Dominican architect and writer Gustavo Moré has pointed out that in the 1930s serious plans were laid to 'create a space for citizenship'. Specifically, the 'first truly republican project', by architects José Antonio Carol and Guido D'Alessandro in 1937, was a plan to develop buildings and public spaces on a northern axis referencing a new Presidential Palace. The southern terminus of this axis was the Caribbean Sea, as Moré notes, a move that likely influenced the layout of the Feria de la Paz (Moré and Martínez, 2005). With the fair's inauguration, Trujillo could now claim to have created a new civic centre, and with this act, he had decisively reordered the hierarchy of Santo Domingo's public spaces. During Trujillo's regime, in this and other projects, we witness a transformation in which the already decentralized capital city's sites of memory grow even more disassociated from each other.

The Feria de la Paz represents Trujillo's commitment to celebrating and memorializing his family's power and his own accomplishments. After all, the fair was planned as a celebration of the twenty-fifth year of the 'triumphant' Trujillo regime. The extent of his influence is undeniable as he was depicted everywhere, even casting a shadow over his brother Hector B. Trujillo y Molina, who was actually the President of the Dominican Republic at the time of the fair. The *New York Times* had reported early in 1955 that Ciudad Trujillo 'enjoyed one of the largest favorable trade balances in its history'.[10] Trujillo was hoping that in addition to increasing the successes with sugar, coffee and tobacco exports, the country would soon enjoy additional financial benefits from the tourism associated with the fair. New hotels, new entertainment facilities, and expanded transportation services, including new modern superhighways, were all ingredients of the tourism development plan aimed at guaranteeing visitors a pleasant experience. Success in the tourism development effort, however, would depend on how the international community viewed Trujillo's administration. 'There are indications that the chief of state is wavering between turning his role into one of benevolence, if such is possible, and continuing as a relentless dictator',[11] as journalist Paul P. Kennedy wrote in the *New York Times*, just three months before the fair opened. This depiction suggests an attempt at a partial, or at least potential, attempt by Trujillo to soften his international image.

On Opening Day, 20 December 1955, relatively few of the exhibitions were complete; only part of the US Pavilion was open and none of the other foreign pavilions were ready. In the United States, opponents of the Trujillo regime protested in New York City outside the offices of the Dominican Information Center. On their placards, they criticized Trujillo and called for a boycott of the fair. The *New York Times* reported that one sign asked how a fair could be dedicated to the '"brotherhood of the free world" in a tyrannical dictatorship'.[12] Far removed from the protestors in the freezing northern city of New York, the fairgrounds, situated on 800,000 square metres along the tropical Caribbean coastline, were bustling with visitors celebrating the opening day. The grounds were bound on the north and south by Avenida Independencia and Avenida George Washington (the *malecón*) respectively, and on the east and west by Calle Héroes de Luperón and Avenida Abraham Lincoln. As an iconographic centrepiece for the exposition, Trujillo and his architect González chose to appropriate a pair of well-known

American icons, the Trylon and Perisphere of the 1939 New York World's Fair that Trujillo had visited. What had been in New York pristine, white abstractions symbolizing a perfect tomorrow (the Perisphere housed the model town Democracity) were transformed in this Caribbean city into militarized tropical signposts that carried a different message. The trylon-esque vertical element now bore the five gold stars that identified Trujillo's military rank of *generalísimo*. It loomed large behind a blue globe that presented the Western Hemisphere, which, when viewed in the reflecting pond on which it sat, inverted the representational globe, making South America appear above North America. Modelling himself after the US marines who trained him, Trujillo was, to a large extent, a product of the United States, and he borrowed and personalized well-established symbols, imbuing them with self-referential meanings. Through this fair, he celebrated his own vision of a 'free world' with rituals and monuments that were highly personal and were meant to elevate his status as a ruler. The exposition was as much a display of Trujillo's personal powers as it was a celebration of the city.

Among the many buildings erected on the fairgrounds were large institutional buildings finished with lavish materials, which now function as the seat of government in the city, and include the Congress, the Supreme Court, the Municipal Palace, multiple government office buildings, and the National Lottery Building. Buildings and exhibitions at the fair were sponsored either by countries, businesses, or Dominican institutions. It was reported that forty-two 'free' (non-Communist) countries participated in the fair, although only the following eighteen countries with exhibits or full-blown pavilions were published in the *Albums de Oro*: Canada, China, Colombia, France, Germany, Guatemala, Holland, Indonesia, Italy, Japan, Mexico, Nicaragua, Panama, Peru, the Vatican's Holy See, Spain, the United States, and Venezuela. Some significantly large structures and smaller pavilions were dedicated to agriculture, the armed forces, the Dominican cement factory, education, finance, foreign relations, hydraulic works, industry, commerce and banking, the Interior Department of Security, Justice and Labour, public health, public works, sugar, and the university. A large fascist-looking building was dedicated to Trujillo's political party, the Dominican Party. The grand Beaux-Arts palace, the Temple of Peace, and the Theatre of Water and Light also graced the fairgrounds.[13]

Spectacular events were showcased almost every evening, not to mention other simultaneous events, such as the *Feria del Libro*, the *Feria Ganadera Nacional*, and an international Baseball Championship. One of the most telling events, which speaks of the self-aggrandizement that this fair truly represented, was the parade, the *Corso Florido*, which took place on 10 April. If the visitor was not able to see every pavilion or the actual city, numerous Trujillo marvels were replicated on floats and paraded through the grounds (figure 10.3). At a world's fair, it was

Figure 10.3. Parade float representing 'Finances' at the *Corso Florido*, 1955.

customary for the host nation to boast with a few domestic pavilions, but in this case, the fairgrounds were a *Ciudad Trujillo* showroom.

The extravagant Feria de la Paz usurped a significant amount of the nation's budget, but also delivered, as promised, many fine buildings, which would eventually serve as the country's administrative and military quarters. Much of the event, though, was a display of superficiality and nepotism. Historian Lauren Derby writes:

> During a full year of trade fairs, dances and performances, the culminating apex of the event was a 'floral promenade' showcasing the dictator's daughter, sixteen-year-old María de los Angeles del Corazón de Jesús Trujillo Martínez (better known as Angelita) who was crowned queen during the central carnival parade. (Derby, 1998, p. 380)

Derby mentions how Angelita was cloaked in a 75 foot garment made of the skins of six hundred Russian ermines (figure 10.4).

The two *Albums* depicted a Dominican population of almost exclusively European descent, with African-descended Dominicans portrayed in only a few places, such as when featuring agricultural products, or pictured while tending horses. The juxtaposition of images, showing the wealth of an elite class within the settings of the island's natural beauty, presented a fabricated impression of the nation and society, the way Trujillo wished it to be remembered. Atkins and Wilson (1998, pp. 78–79) point out that Trujillo reinforced an exclusively Spanish heritage that was Roman Catholic and racially pure by 'stressing the fear of the Ethiopianization' and miscegenation; he thus actively sought 'to whiten' (*blanquear*) the population – which included encouraging European, and severely restricting Haitian, immigration.[14] Many historians have since revealed a fact that Trujillo unsuccessfully attempted to obscure: that he was partially Haitian himself.

The Feria de la Paz was the site of Trujillo's last attempt at focusing attention

instantes después de haber colocado la corona
real sobre la linda cabeza de Angelita I, el señor
Presidente Trujillo escucha los aplausos de la multitud congregada en el Teatro Agua Luz.

Figure 10.4. Angelita's coronation with Héctor B. Trujillo. (*Source*: *Album de Oro*, Tomo I, 1956)

on his power, and there was no question that in doing so he had exhausted and depleted the nation's wealth (when Trujillo was assassinated, his family had almost $300 million in accounts outside the country) (Moya Pons, 1995, p. 375). The *New York Times* first reported on the Feria de la Paz in 1954 in a very positive tone, but this changed while the fair was in progress, with mixed reviews and increasing accounts of Trujillo's tyranny being published in the American press. Trujillo was finally caught in the bright spotlight of international disapproval, and he spent the last years of his life in condemnation, surviving as many as seven attempts on his life before finally succumbing to assassins' bullets.

Hegemony and Contestation in Public Space

But these architectural gestures were only the most visible layer of a more deeply ingrained and a much more sinister system of influence imposed on Dominican society. Wiarda (1970, p. 33) calls this 'the field of totalitarian thought control with even tighter supervision and more clever manipulation of all communications media, education, and intellectual life'. He considers this type of thought control a new phenomenon of modern totalitarianism that was the product of twentieth-century developments in the technology of communications (*Ibid.*, p. 124). Antoni Gramsci's (1891–1937) concept of 'hegemony' is applicable here to describe the domination of one class or nation, in this case one person, over an entire population.[15] Perhaps more extreme than in other cases of hegemonic influence, in the Dominican Republic, a pervasive civic language of place-names and slogans led to intrusions into the most private of spaces: the home. An example of this type of forced allegiance is depicted in the movie *In the Time of the Butterflies* (2003), which tells the story of the Mirabal sisters, who were murdered under Trujillo's regime. The movie is based on the 1994 novel of the same name by Dominican-born writer Julia Alvarez. Orwellian scenes in the film portray typical domestic life as a family is watched over by the ubiquitous portrait on the wall of Trujillo in a Christ-like representation. In the days of the totalitarian regime, the absence of such a portrait hanging in one's living room was cause for serious suspicion. With it, family life included the requisite if tacit acknowledgement of Trujillo's regime. Here the megalomaniac met the ordinary citizen on a daily basis. Another novel, *The Feast of the Goat* (1999) by Peruvian writer Mario Vargas Llosa, presents a more explicit account of the sexualized and rage-filled autocracy founded on the Latin American machismo. Public response to this severity and pervasiveness of authoritarianism took many forms. In an extreme case, Vice-President Jacinto Peynado praised the dictator by installing a neon sign at his front door that stated 'God and Trujillo' (Roorda, 1998, p. 96). Trujillo's transformation of the nation into his own personal shrine was a reflection of the flamboyance that surrounded his participation in official governmental and historical celebrations and military rituals. Roorda claims that at the time of Trujillo's assassination, some 1,800 sculptures sat in public squares and buildings throughout Ciudad Trujillo, 'one for every ten square miles' (*Ibid.*, p. 97), while Atkins and Wilson (1998, p. 66) put that number at exactly 1,870 monuments.

During Trujillo's reign, rioting and protests were strictly prohibited and punishable by death; only during the civil war that followed the dictator's fall did the streets of Santo Domingo see significant riots (figure 10.5).[16] Decades of controls on public behaviour and the absence of a collective history of mass public demonstration meant that as far as collective memory is concerned, Dominicans have not used public space for political purposes as in other Latin American cities. Unlike Latin America's well-known plazas of public protest and anti-government expression, such as those in Bogotá, Buenos Aires, and Mexico City, for example, Santo Domingo's major public spaces have only a few times been the scene of large-scale demonstrations or lengthy protests.[17] During a period after World War II, when Trujillo's regime put on a façade of democratic reform, trade unions were allowed to organize and the press was allowed to criticize Trujillo. But this loosening of the leash of autocracy was short-lived, and once Trujillo was officially elected, all opposition to his government was squelched (Roorda, 1998, p. 33). The public spaces where pro- and subsequently anti-Trujillo events were staged but do not appear on the typical tourist map are, nevertheless, telling of political struggles for a local community, although they are spaces that have been referred to as both fragmented and isolated by the local professional community (Ramírez *et al.*, 2005, p. 103).

The Un-making of the Goat's Shrine

On 30 May 1961, Trujillo was shot dead by a group of men who had followed him from the fairgrounds down the *malecón*. Six months later many references to Trujillo in the city – statues, streets, buildings and plazas – praising his supposed heroism and generosity, were unceremoniously removed. This action was taken by both vehement anti-Trujillistas and by the surviving elements of the government,

NACE LA LIBERTAD

La jubilosa multitud vitorea el nacimiento de la libertad en Santo Domingo. Alborozo, gritos y palmetas se aunaron para formar el carnaval de la libertad dominicana.

Figure 10.5. *'Nace la Libertad'*, *Unión Cívica*, 25 November 1961, p. 5. (*Source*: The Nettie Lee Benson Latin American Collection, University of Texas at Austin)

who had significant motivation to appear reform-minded.[18] G. Pope Atkins and Larman Wilson (1998, p. 145) note how, a few months after the assassination, students of the Universidad Autónoma de Santo Domingo 'rebelled, destroyed statues and other symbols of Trujillo's megalomania'.

Unión Cívica provides an excellent chronicle of the state of affairs during the momentous year of 1961, when the Trujillo regime was gradually toppled. The newspaper presented a forum where democracy was discussed both in prose and through graphic visualizations. More importantly, it instructed Dominicans in the acts of erasure, demonstrating the many ways that the city was being and could be transformed and reclaimed. Certainly, it was an instrument of partisan politics, and many have pointed to its biased support of a conservative, elite candidate, but nevertheless it served as a means to publicize the atrocities that had previously been discussed only in private.

Step by step, Santo Domingo was reclaimed. In 1965, 4 years after Trujillo's demise, the fairgrounds were converted into a memorial commemorating in general all those who died under his regime, and, specifically, those whose assassination attempts failed in the historic event that took place in 1949. Photos of these heroes were mounted where the emblems of the nations who participated at the fair were displayed (figure 10.6). The relatives of the people the memorial honoured spearheaded the effort; they wanted to memorialize the political exiles that plotted Trujillo's demise. The entire fairgrounds were renamed The Centre of the Heroes of Constanza, Maimón and Estero Hondo to honour these Dominicans who suffered retribution in this failed attempt to overthrow Trujillo. The project to mount photos of the exiles on the walls that flank the fairgrounds' main plaza, locally referred to as '*la bolita del mundo*' (the little ball of the world), was organized by the mothers and wives of these men. After the assassination, public meetings were held between 1961 and 1963 to publicize the atrocities, many covered by the *Unión Cívica*.

Understandably, the remodelled fairgrounds went unnoticed when they were used in Frances Ford Coppola's movie *The Godfather II*, which was filmed in 1972. The main plaza set the stage for a downtown New Year's Eve celebration in Havana at the moment of the Cuban revolution. With the removal of the five gold stars that marked Trujillo's military rank – *generalísimo* – from the fair's obelisk, and the mounting of the images of *los Héroes*, only the pink and blue paint remained as a symbolic reminder of the Feria and Trujillo's military heyday. Currently abandoned and deteriorating, the plaza has been a haven for the city's poor and prostitutes. It is a reminder for some of a time of great civic works, and for others, of the many who lost their lives during the years of the bloody Trujillo dictatorship.

In 1966, a second monument, the Monument Mausoleum to the Heroes of

Figure 10.6. Two views
of the Mausoleum gate
designed by Domingo
Liz. (*Photos*: Robert
Alexander González)

Constanza, Maimón and Estero Hondo, was completed on an adjacent site. The mausoleum designed by local sculptor Domingo Liz (1931–) houses the remains of the same men whose pictures were hung around the plaza. However, in its architectural expression, the mausoleum represents an antithesis of the plaza's original aim. It transformed what might have been remembered as a glittering public space, dampening its celebratory aesthetic with sombre greys and dark tones. The grim nature of the dictatorship is first captured in the pair of angels seeking solace from each other at the entry gate. From the streetscape, there is little to suggest that one is approaching the mausoleum. There is no urban promenade, street furniture or other indications of the way the mausoleum, and hence its presence is kept uncelebrated. Liz was responsible for six civic projects in Trujillo's reign, including the military academy, the military club and some municipal palaces, but he says that none were destroyed in 1961 because he never included references to Trujillo in these public works.[19]

Paralleling the tension that exists between these two adjacent sites of memory, we find another more explicit duality played out on the *malecón*. Two prominently situated monuments capture the deep symbolism that has pervaded the distinct form of memorializing in Santo Domingo. They are obelisks, one 'male' and one 'female' – as Dominicans like to refer to them. The female obelisk was erected when Trujillo paid the nation's external debt; the monument was originally called the Treaty Trujillo-Hull Monument, after the treaty that was ratified on 15 February 1941. It was designed by Spanish exile, Tomás Auñon and was 'the object of an enormous government propaganda campaign to make Trujillo appear as the restorer of the country's financial independence' (Moya Pons, 1995, p. 366). It is irreverently referred to as a female monument because some locals claim it looks like two legs that spread outward to the sky. The male obelisk is a single, phallic column. It commemorates the commencement of the Era de Trujillo in 1930. It is the 'male' obelisk that the citizens of Santo Domingo have tried to transform; it has been re-appropriated as a homage to the Mirabal heroines, who played an important part in the regime's downfall (figure 10.7). In 1961, the article 'The obelisk should be destroyed' written by Octavio Mata Vargas for the *Unión Cívica*, called for the destruction of the monument. Vargas described it as a structure that was built to satisfy the vanity of the despot, as an artefact that had no public use, and as a reminder of the regime's cruelty and inhumanity. At the time, there must have been a proposal to have the names of 'the disappeared' heroes engraved on it because Vargas notes that it would be a grave insult to take up this idea. The question of memory was raised as well when he wondered if, in the future, anyone would take up the task of reminding others of the tyranny that took place in the Dominican Republic. Vargas believed that the best solution was to do away with it once and for all, leaving no trace even of the foundation, as though it had

Figure 10.7. *El Obelisco Macho* transformed into the Mirabal Obelisk. (*Source*: *Ciudad del Ozama*, Eugenio Pérez Montás)

never existed.[20] The citizens of Santo Domingo, however, chose to transform the obelisk years later, incorporating the story of the three women who perished.

Known as *Las Mariposas*, Minerva, Patria, and María Teresa 'Mate' Mirabal were killed on 25 November 1960, after years of working with the underground '14th of June' resistance movement aiming to depose Trujillo. Their death led to a significant decrease in popular support for the dictator, a strengthening of the resistance movement, and a more open disapproval from the Catholic Church. Yet it took nearly four decades for *Las Mariposas* to be publicly memorialized in Santo Domingo. On 8 March 1997, the city unveiled a mural depicting the Mirabal sisters, called *Un Canto de Libertad* (A Song of Liberty), which was painted on the 137-foot 'male' obelisk that Trujillo had erected to honour himself. The now transformed obelisk eulogizes the many Dominican women and men for liberty, and narrates their struggles. Today, for some who stroll or drive down the *malecón*, these urban landmarks are merely a sign of femininity and masculinity, remnants of Trujillo. For others they represent redemption for those oppressed by the regime and of progress towards democratic ideals.

Still other groups in Santo Domingo have revisited the country's gruesome past and attempted to reclaim the nation's cultural landscape. In 1998, the Heroes of May 30 Foundation, a group honouring the men who actually got to Trujillo on that date in 1961, constructed their own monument on that very highway, Autopista 30 de Mayo, where Trujillo was followed and ambushed in his light blue 1957 Chevrolet Belair. This monument, called the Monument to Justice, Heroes of May 30, is the work of Silvano Lora (1931–2003), a local sculptor muralist, writer and revolutionary. The symbolically charged black and gold-tiled monument depicts an oppressive face with two hands extending out, facing an isolated set of flames burning a wrapped female body, which is held up by the flames. Located some 2km south-west of the city centre, the sculpture brings to mind a sacrificial offering, and the outstretched hands, a warning to future generations. Lora, once described as the Dominican Quixote, was a colleague and fellow patriot of Domingo Liz, and through this friendship, both monument and mausoleum share a place in the history of revolutionary art that emerged in Trujillo's era.

In 1992, another major monument was finally completed; the Columbus Memorial Lighthouse was built to celebrate the quincentenary of the 'discovery' of America. The monument building activity surrounding this memorial is further complicated by another political figure, Joaquín Balaguer (1907–2002), who succeeded Trujillo. Appointed President of the Dominican Republic six times, Balaguer was considered by many untrustworthy and a continuation of the Trujillo regime. He was in office when Trujillo was murdered and, having served as Vice President (1957–1960) he assumed the presidency when the dictator's brother, Hector Trujillo, resigned. A *trujillista* himself, his involvement in future monument building often tainted the meaning ascribed to these new monuments. He suffered public condemnation after requiring that the building of the Lighthouse include a wall – dubbed the 'wall of shame' in the media – which was meant to hide from view the adjacent neighbourhoods that lay in squalor. Balaguer was regarded as no different from Trujillo; in fact, some claim that more bloodshed occurred under his tyranny.

Frameworks of power and citizenship in Santo Domingo changed radically after Trujillo's demise, but with the lighthouse, one could say that the forces of days gone by prevailed. The lighthouse reinforces the Christianizing effort that Columbus ordered upon his arrival – it shines an enormous cross onto the night sky. It also represents both an early twentieth-century attempt by the Pan-American Union to control hemispheric discourse in the Americas as well as the United States' imposition of a custom's receivership on the small nation. These goals can be directly traced back to the Monroe Doctrine of 1823.[21] Moreover, the monument has also been criticized for the huge electricity bill with which it saddles the public finances; this is attributed to Balaguer's administration and has led to a feeling of resentment towards the monument as well as to its general obsolescence in local opinion.

The Dominican Republic experienced one of the more lengthy and severe totalitarian regimes in Latin America, and its capital city has withstood a wrenching, continual process of civic change and self-definition. The marks of Trujillo's regime will not easily be completely erased. Any sign of regression awakens old ghosts. Recently, controversy arose when it was mis-communicated in the press that the Dominican Republic's President Hipólito Mejía wanted to rename Pico Duarte, the highest mountain in the Antilles, Pico Trujillo. This information, before it was known to be erroneous, immediately drew criticism in the press. Apparently, when the legislative branch passed a new bill on National Parks it referred to an old law dating back to Trujillo's era, when the highest peak in the Caribbean was named after Trujillo. Before and after Trujillo's rule, this natural landmark was named Pico Duarte. It is not surprising to find that in the popular website www.dr1.com, where people can log on anonymously, the occasional pro-*trujillista* appears, inciting discussions on the merits of Trujillo's regime, compared with the more recent ones.[22] This is a telling account of the freedom of expression that the Internet now enables. When anonymity is not available, however, protesting the *status quo* might take the form of agreeing to forget altogether or leaving the subject of the past at the door.

Public Expressions and Silenced Heroes

Since 1961 five organizations have formally participated in the memorial building process: the *Fundación de los Héroes de Constanza, Maimón, y Estero Hondo*; the *Fundación de los Héroes de 30 de Mayo*; those organized in remembrance of the Mirabal sisters; the nation's armed forces;[23] and those who have organized the recent competition to renovate the fairgrounds. The nation itself contains many more factions that are just anti-Trujillo or pro-Trujillo, or that hold him in high regard for his efforts to transform the nation and capital city. In this sense, Moya Pons refers to the enormous bibliography of apologist literature, admittedly mostly written during Trujillo's reign. Despite such strong viewpoints about the past, the monuments examined in this chapter denote extraordinary events, but are still treated as ordinary, uneventful spaces by the average citizen. The old fairground plaza is especially shunned because it has become the centre of the city's growing sex trade and has been raised as a serious problem in the press. Only the cases of extreme controversy about historical events seem to enliven the city in public debate. The tourist industry's shunning of these monumental destinations reflects the status of a country that is still, if only silently, grappling with its troubled history.

Because Trujillo's regime was so pervasive throughout the nation, it practically suppressed all public expressions of discontent. Unlike in other Latin American cities, no protest movement of any significance was able to flourish. Advances in technology and the media allowed public spaces to be used to develop and perpetuate total control of the kind witnessed in Trujillo's regime, which was the longest of any in Latin America (Wiarda, 1970, p. 1). For nearly three decades, there was absolutely no tolerance of opposition. Even silence, interpreted as neutrality, was a crime. Max Frankel wrote: 'Fear stalks the Dominican capital. Missing are all dignity and choice; there is not only no freedom to speak, but no freedom to remain silent'.[24] Plans for the eventual revocation of this regime had to be made under conditions of extreme secrecy. That the silent butterflies, *las mariposas*, were symbols of the resistance suggests the level of fear that rendered the nation voiceless for so many years.

The fairground's main plaza, *la bolita del mundo*, is soon to be unsilenced; it is predicted to be the next site of transformation. In June 2004, the city announced that a new municipal centre was in the final planning phases. City officials renovated the fair's plaza in May that year, a task that involved the removal of the photographs of the heroes. With the support of the local government, the Association of the *Centro de los Heróes*, the group that oversees the plaza, commissioned local architect and urban planner, Marcos Barinas, to organize a competition to restore and re-envision the fairgrounds. One must ask, to what extent did the removal of the photographs pre-empt the 'visioning' exercise that the competition was supposed to achieve? The international competition was not very fruitful; it only yielded eight entries. Coincidentally, local architect, Gustavo

Moré, who has done extensive work adjacent to the plaza, and Puerto Rican Emilio Martínez shared the first prize with another team; Moré had just completed the adjacent Supreme Court of Justice and the Attorney General's Office. Moré and Martínez write that these government structures, which Moré designed, give 'this space a remarkable meaning regarding citizenship' and that the high quality of the fairground scheme developed by Guillermo González invites the possibility of a new national civic centre, 'a new agora for the future that can admit the notion of contemporary and democratic "Dominicanhood" and that can realize the meaning of a nationality in permanent transformation that demands stable referents in time and space in order to consolidate its own identity' (Moré and Martínez, 2005, p. 95). It is unfortunate that their interest in transforming the fairgrounds into a 'new, efficient, modern, global space capable of representing nationhood' – something that is not clearly defined in their design brief – did not have room for the photographs of the deceased. Their notion of nationhood can only be discerned from one of the design interventions, where it is noted that the verses of the National Anthem would be etched into limestone in the Plaza of the Constitution, which they call a new 'citizenship space'. The Mausoleum would remain, but they recommend that it be redesigned via a national design competition, which would, in their words, 'legitimize it both historically and socially' (*Ibid.*, p. 99).

When, in Iraq, statues of Saddam Hussein were pulled down, the world experienced the *déjà-vu* of towering statues being toppled, reminiscent of the falling statues of leading public figures of eastern Europe and the Soviet Union. The unique case of Santo Domingo, where remnants of the Trujillo regime include not only monuments but the urban icons that give the city its signature cityscape – the fairgrounds and the *malecón*'s obelisks – presents a condition where complete erasure will probably never occur. The erasure that did take place in 1961 in Santo Domingo and the nation did so with much fervour, but any reference to one of these monuments, any act that revisits the meaning behind these icons, is in itself a revisitation of Trujillo's legacy. It is a recognition that pulls at the frayed edges of democracy in the nation and contemporary designers know to approach the plaza carefully.

Yet ordinary citizens' memories and experiences are not celebrated with ritual or ephemera, as we see occur with the weekly marches of Mothers of the Plaza de Mayo in Buenos Aires, who can publicly honour the disappeared. Many Dominicans just want to move on, and many have. Some will still say that there is no consensus on how to evaluate contentions around the legacy of the fallen regime: the plights of human misery, and the building of the infrastructure that transformed the city and the nation – two legacies that many others would contend should not be weighed against each other. Dominicans are suffering the plight of the child begotten from the act of rape. Trujillo gave the city a new 'modern' life, but the path to this end was gruesome. Many of the city's public spaces preserve the memory of very few but very memorable extraordinary events, and over time, understandably, they have become ordinary places, some even neglected and forgotten.

Notes

1. Trujillo was President during the following periods 1930–1934, 1934–1938, 1942–1947 and 1947–1952. He was also the foreign minister from 1953–1961.
2. *Unión Cívica* was produced by National Civic Union, one of the three political parties that opposed Joaquín Balaguer after Trujillo's regime ended. The other two parties were the Dominican Revolutionary Party and the June the 14th.
3. *Unión Cívica*, Año 1, No. 32, 25 November 1961.
4. The competition organizers decided at the outset that three jurors would be chosen, each one representing Europe, Latin America, and North America. The Memorial Lighthouse that was finally built by local architect Juvenal Carbonell bears little resemblance in detail to the original Gleave design, but the overall composition remained the same.
5. The lighthouse was supposed to be funded by the American Republics with funds according to population. The Dominican Republic and the United States had already agreed to each pay $300,000. When questionable election results placed Trujillo in office, the *Listín Diario* reported that the US Congress would vote against the appropriation for the monument. (Pulliam to Rowe, 1 April 1930, Columbus Memorial Library Archives).
6. The cheque included the paid quotas from the Dominican Republic, Honduras, Nicaragua, and Panama. (Memorandum for Dr. Manger from Mr. Curtiss, 14 March 1955, Columbus Memorial Library Archives).
7. *La Arquitectura Dominicana en la era de Trujillo. Album 1* (Ciudad Trujillo: Impressora Dominicana, 1949). For a review of the architecture of the era of Trujillo see Gustavo Moré's chapter titled, 'History,' in Calventi (1986).
8. The engineer-architects included: Andrés de los Santos Báez, Antonio Molina (lieutenant), Petrus Manzano Bonilla and Rafael Félix (professions not noted), Juan Isidro Prandy Báez (helper), Mauricio Estrella (architect-engineer), José C. Farías, MI Francisco, B. Batista (students), and Raúl Cordero (auditor).
9. Considered the father of modern architecture in the Dominican Republic, González began his training as an architect in the Public Works office in Santo Domingo between 1916 and 1924 during the US occupation. Graduating from Yale, he travelled to Europe and returned to his native country in 1936. He designed the famous Hotel Jaragua, which was demolished in 1985.
10. *New York Times*, 5 January 1955.
11. *New York Times*, 14 September 1955.
12. *New York Times*, 21 December 1955, p. 51.
13. This building has been remodelled twice since its erection in 1955, first in 1989 and then in 1993 by President Joaquín Balaguer. It was featured in the movie, *The Godfather II*.
14. Trujillo is also remembered for ordering the massacre of Haitians in October 1937, where as many as 20,000 unarmed men, women, and children were slaughtered.
15. Antoni Gramsci, *Selections from the Prison Notebooks* (Lawrence and Wishart: London, 1971). One *New York Times* story published on 14 September 1955, reported that: 'Generalissimo [sic] Trujillo and his family are the wealthiest people in the republic and are among the largest property holders… The chief of state owns wholly or with his family the leading hotels, shipping, radio and television facilities, sugar mills, cement works, automobile agencies, banks, brewery and other ventures'.
16. John Barlow Martin, in *Overtaken by Events*, writes about the civil war rioting that occurred and how detectives from the Los Angeles Police Department were brought to Santo Domingo to train the police in riot control. Howard J. Wiarda (1970) writes that during Trujillo's re-election campaign for his second term, minor revolts broke out in the provinces against the government. 'They were quickly crushed and this kind of

uprising soon ceased to occur altogether'. Wiarda goes on to say how during Trujillo's fourth term, public demonstrations were prohibited with even greater force. The author also notes that only one major strike occurred in Trujillo's time, in January 1946, when the sugar workers demanded higher wages.

17. See chapters on these cities in this book, and Montilla (2003).
18. As Atkins and Wilson point out, Trujillo's tributes were not immediately erased. In November 1961 some members of the Trujillo family attempted to regain control of the country and Trujillo's hand-picked president, Balaguer, was considered politically unacceptable. They write, 'Disorders and violence mounted, aimed at dislodging Balaguer, who declared a state of emergency. At this time the removal began of virtually all statues of the slain dictator throughout the country' (Atkins and Wilson, 1998, p. 126).
19. Interview with Domingo Liz, 2 July 2005.
20. *Unión Cívica*, 31 January 1962.
21. It was the Custom's Receivership, William Pullman, who while stationed in Santo Domingo, worked with the Pan-American Union to coordinate and facilitate this competition. With this, the memorial is as much a product of the history of US dominance in the Caribbean as it is a symbol of an organization wanting to bring the Americas together. The Monroe Doctrine was an American diplomatic decision, initiated by President James Monroe, which aimed to limit European expansion in the Western Hemisphere, the United States assuming the position of 'protector'.
22. www.dr1.com/forums/archive/index.php, 22, 24 October and 3 November 2003.
23. To put into perspective the currency with which this monumentalizing activity is taking place, in April 2003, another monument was built in the mountainous area near the town of Constanza. With what may be interpreted as the hope of vindication, the nation's armed forces decided to honour the men who died defending Trujillo in June 1959, the failed attempt on his life, which was later memorialized with photos on the fair's main plaza and with the adjacent mausoleum. The monument was built in the form of a plaza. Its distance from the city was no consolation to those opposed to it, though, and it caused much public debate, including a call for its destruction. President Hipólito Mejía eventually ordered the Armed Forces Minister José Miguel Soto Jiménez to rededicate the monument, to commemorate instead the founding father of the city, Juan Pablo Duarte. Journalist Eduardo Díaz captured the sentiment of many when he said that the monument constitutes an embarrassment because it honours obedience to crime and attempts to justify the actions of those who were in service to the dictator. The armed forces responded by saying that 'the original concept of the monument was apparently aimed at avoiding "misunderstanding and suspicion" directed at the armed forces, as well as healing "injured sensibilities"'. A relative of the Mirabal sisters, Minou Tavárez Mirabal, also publicly denounced the monument.
24. *New York Times*, Sec. 4, 4 June, p. 4, quoted in Wiarda (1970, p. 55).

References

Atkins, G. Pope and Wilson, Larman C. (1998) *The Dominican Republic and the United States, From Imperialism to Transnationalism*. Athens, GA: The University of Georgia Press.

Boyer, M. Christine (1998) *The City of Collective Memory: Its Historical Imagery and Architectural Entertainments*. Cambridge, MA: MIT Press.

Calventi, Rafael (1986) *Arquitectura Contemporánea en República Dominicana*. Santo Domingo: Editora Amigo del Hogar.

Derby, Lauren Hutchinson (1998) The Magic of Modernity: Dictatorship and Civic Culture in the Dominican Republic, 1916–1962. PhD thesis, University of Chicago.

Gazón Bona, Henry (1949) *La Arquitectura Dominicana en la Era de Trujillo. Album 1*. Ciudad Trujillo: Impresora Dominicana.

Gramsci, Antoni (1971) *Selections from the Prison Notebooks*. London: Lawrence and Wishart.

Martin, John Barlow (1966) *Overtaken by Events: The Dominican Crisis from the Fall of Trujillo to the Civil War*. New York, NY: Doubleday.

Montilla, Armando (2003) Taking the streets: Caracas 2002–2003. *Aula*, 4, pp. 44–53.

Moré, Gustavo Luis and Martínez, Emilio (2005) The space of Dominicanhood: constitutional representativity. *AAA* 020.

Moya Pons, Fran (1995) *The Dominican Republic. A National History*. New Rochelle, NY: Hispaniola Books.

Palm, Erwin Walter (1974) *Arquitectura y arte colonial en Santo Domingo*. Santo Domingo: Editora de la Universidad Autonoma de Santo Domingo.

Pérez Montás, Eugenio (1998) *Ciudad del Ozama: 500 años de historia urbana*. Barcelona: Patronato de la Ciudad Colonial de Santo Domingo.

Pieiter, Leoncio (1958) *Ciudad Trujillo*. Ciudad Trujillo, D.N.

Ramírez, Arístedes and Despradel, Arturo, Valdez, Michelle, Valdez, Gustavo and Pons, Daniel (2005) Another part of the city. *AAA* 020.

Roorda, Eric Paul (1998) *The Dictator Next Door, The Good Neighbor Policy and the Trujillo Regime in the Dominican Republic, 1930–1945*. Durham, NC: Duke University Press.

Walker, Stanley (1956) *Generalissimo Rafael L. Trujillo*. New York, NY: The Caribbean Library.

Wiarda, Howard J. (1970) *Dictatorship and Development, The Methods of Control in Trujillo's Dominican Republic*. Gainesville, FL: University of Florida Press.

Index

DATE DUE